The North Carolina State Fair

Nighttime view of the Ferris wheel at the 1996 fair. Photo courtesy of NCDA&CS.

The North Carolina State Fair

The First 150 Years

Melton A. McLaurin

Illustrations selected by Paul Blankinship

OFFICE OF ARCHIVES AND HISTORY
NORTH CAROLINA DEPARTMENT OF CULTURAL RESOURCES, RALEIGH

NORTH CAROLINA STATE FAIR DIVISION
DEPARTMENT OF AGRICULTURE AND CONSUMER SERVICES, RALEIGH

Contents

Foreword

The North Carolina State Fair is a magical experience of sights, sounds, smells, and senses. Nothing in this world is quite like it; everything melds together to create an unforgettable experience. The coolness of the time of year, the motion of the people and rides, the excitement, the smell of the food, and the combination of so many things in one area make the fair a unique environment unlike any other.

The North Carolina State Fair is an American agricultural fair, born from the same spirit that launched America's first fair in Massachusetts in 1810. Agricultural fairs were entering their golden age when the North Carolina State Agricultural Society inaugurated its fair in October 1853, beginning a grand tradition that brought the people of the state together in Raleigh each autumn for education and entertainment. In that year Franklin Pierce became the fourteenth president of the thirty-one-state American Union. Westward expansion, improved agricultural techniques, and the railroad were changing America.

Over its 150 years, the fair has taken place at three different sites in Raleigh. It has been held at its current site for seventy-five years, and for thirty of them I have been actively involved with the fair. I started working for the fair in 1973. My work involved cleaning the horse stalls and maintaining the landscape. I was promoted many times over the years and in 1997 was named fair manager. I remain committed to improving and maintaining the grounds.

The fair manager has an important role in shaping a vision for the fair. J. S. Dorton led the fair into the modern, post-World War II era. Art Pitzer oversaw the completion of several exhibit buildings and the development of the off-season rental program. Sam Rand's focus was on the campus infrastructure. I have been fortunate to build upon the work of these men in the improvement of the fair and the fairgrounds proper. As the current manager, I am proud to have been involved in the renovation of the most historical and unique buildings on the grounds: the Commercial and Education Buildings and J. S. Dorton Arena. These buildings add to the magic and uniqueness of the fair and are treasures to the state.

With an annual attendance of 700,000, the state fair is North Carolina's largest event and is the largest ten-day agricultural fair in the nation. Just as it did 150 years ago, the fair still serves as the annual meeting place for people from all over the state. It is a chance to look at the handiwork in the Commercial and Education Building, the expansive art competition in the Gov. Kerr Scott Building, the livestock in the Graham Buildings, and the beautiful house plants, cut flowers, and gardens at the flower show. The state fair awards more than $500,000 in prize money at these various competitions.

The fair is a timeless event that will be relevant 150 years from now. Its ability to adapt to current community needs will keep it a vital part of the state's history for many years to come. Competition, the role of agriculture, and the types of exhibits may change, but the need to learn and be a part of one's community will remain the same. We, the staff of the State Fair Division, remain committed to the agricultural roots of the fair. The fair hosts one of the largest all-breed horse shows in the nation and has strong livestock programs and innovative educational exhibits focused on asking that I excuse the unusual clutter, the result of preparations to vacate his office. For a few moments he engaged me in the down-home small talk of two North Carolinians with a professional interest in one of the state's great institutions—a strategy designed, I suspected, to test whether my credentials as a Tar Heel were genuine and if my knowledge of the fair passed muster. I dropped the name of a friend of my father's, a tobacco farmer long active in the North Carolina Farm Bureau, who, like my father, was now in his eighties. Graham immediately recognized the name and astutely positioned me among the families of rural and small-town eastern Cumberland County.

With my credentials established, I moved into the interview, asking him if the story I had heard about his meeting his wife at the state fair were true.

"Yes, it is," he smiled, as if taking pleasure in both his memory and my knowledge of the event.

"I wonder if we could begin the interview by you telling me about that meeting?" I asked, and he was off and running.

"It was in 1941. I remember it well. My wife-to-be, I met her at the North Carolina State Fair, at the waterfall, a little wildlife exhibit there, waterfowl and so forth. She was teaching home economics in Rowan County, that being my home, and I had run into her, but I never had really met her. She brought a group of children, her students, down here to the state fair, and I made arrangements to be out there, and happened to see her. I hitch-hiked a ride back on the school bus, and I infuriated the little fella that was sitting with her, that came down with her, 'cause I took his seat going back. His name was Frank White; I remember it well. Anyway, we met there, and that was the beginning of a romance and then my wedding and my wife of fifty-four years. So, a lot of things can happen at the state fair and will happen."

And so they have, to the delight of generations of North Carolinians who have made the state fair North Carolina's largest and most beloved continuing social event. For a century and a half the North Carolina State Fair has educated, entertained, and enthralled North Carolinians from Manteo to Murphy, from West Jefferson to Wilmington. The state fair began as an effort on the part of antebellum North Carolina's leading planters and agricultural reformers to convince the state's

During Graham's long tenure in office, the state fair changed dramatically. Graham oversaw the construction of most major exhibit halls, the modernization of the fairgrounds' infrastructure, and the initiation of year-round use of the fair's facilities. Here Graham speaks to a child at the 2000 fair, his last as commissioner of agriculture and consumer services. Photo courtesy of NCDA&CS.

Foreword

The North Carolina State Fair is a magical experience of sights, sounds, smells, and senses. Nothing in this world is quite like it; everything melds together to create an unforgettable experience. The coolness of the time of year, the motion of the people and rides, the excitement, the smell of the food, and the combination of so many things in one area make the fair a unique environment unlike any other.

The North Carolina State Fair is an American agricultural fair, born from the same spirit that launched America's first fair in Massachusetts in 1810. Agricultural fairs were entering their golden age when the North Carolina State Agricultural Society inaugurated its fair in October 1853, beginning a grand tradition that brought the people of the state together in Raleigh each autumn for education and entertainment. In that year Franklin Pierce became the fourteenth president of the thirty-one-state American Union. Westward expansion, improved agricultural techniques, and the railroad were changing America.

Over its 150 years, the fair has taken place at three different sites in Raleigh. It has been held at its current site for seventy-five years, and for thirty of them I have been actively involved with the fair. I started working for the fair in 1973. My work involved cleaning the horse stalls and maintaining the landscape. I was promoted many times over the years and in 1997 was named fair manager. I remain committed to improving and maintaining the grounds.

The fair manager has an important role in shaping a vision for the fair. J. S. Dorton led the fair into the modern, post-World War II era. Art Pitzer oversaw the completion of several exhibit buildings and the development of the off-season rental program. Sam Rand's focus was on the campus infrastructure. I have been fortunate to build upon the work of these men in the improvement of the fair and the fairgrounds proper. As the current manager, I am proud to have been involved in the renovation of the most historical and unique buildings on the grounds: the Commercial and Education Buildings and J. S. Dorton Arena. These buildings add to the magic and uniqueness of the fair and are treasures to the state.

With an annual attendance of 700,000, the state fair is North Carolina's largest event and is the largest ten-day agricultural fair in the nation. Just as it did 150 years ago, the fair still serves as the annual meeting place for people from all over the state. It is a chance to look at the handiwork in the Commercial and Education Building, the expansive art competition in the Gov. Kerr Scott Building, the livestock in the Graham Buildings, and the beautiful house plants, cut flowers, and gardens at the flower show. The state fair awards more than $500,000 in prize money at these various competitions.

The fair is a timeless event that will be relevant 150 years from now. Its ability to adapt to current community needs will keep it a vital part of the state's history for many years to come. Competition, the role of agriculture, and the types of exhibits may change, but the need to learn and be a part of one's community will remain the same. We, the staff of the State Fair Division, remain committed to the agricultural roots of the fair. The fair hosts one of the largest all-breed horse shows in the nation and has strong livestock programs and innovative educational exhibits focused on advances in agriculture. Agriculture will always have a presence at the fair. Parents will want to show their children on how food is produced, children and adults will want to compete for blue ribbons, and others will come for the sense of community.

Exhibits and the educational aspect of the fair are what separate it from other forms of entertainment such as theme parks. The fair represents an opportunity for the agricultural community to showcase its crops and livestock and, most importantly, an opportunity for the state's ever-increasing urban population to learn about agriculture, North Carolina's leading industry, which generates more than $62.6 billion annually.

The roots of the state fair are deeply embedded in agriculture, and the farmers of this state have benefited from what they have learned by attending the fair. Just as many changes have occurred over the 150 years of the state fair, significant changes have transpired in North Carolina agriculture. In 1853, agriculture in North Carolina was unsophisticated, with poor crop yields resulting from a lack of knowledge about fertilizer techniques,

crop rotation, and pest control. North Carolina's farmers presently use global positioning systems to manage fertilizer applications, many of the worst crop pests have been eliminated or controlled, and a number of farmers market their crops via the Internet. Many of the technologies that have benefited farmers were first showcased at the state fair. I wish I had a crystal ball to see what the next 150 years will bring to agriculture and the fair. I am sure it will involve evolution and progress, as it has since the beginning.

This book, the most comprehensive history of the North Carolina State Fair, surveys the evolution of the annual autumn event through its 150-year history. I hope North Carolinians will enjoy the history and the illustrations it contains. The North Carolina State Fair is rich in heritage and culture, and I hope that you, your family, and your friends will enjoy it for many years to come. See you at the fair!

WESLEY WYATT
Manager, North Carolina State Fair
June 2003

Acknowledgments

As readers of acknowledgment pages know, every book is the product of the efforts of numerous people. Readers are informed of this in acknowledgment pages because it is true, and because authors are grateful for the help they have received in bringing their book to print. This is especially true of illustrated works, and even truer of works such as this, which are illustrated with materials gleaned from several sources and in a variety of formats. The illustration process involves a series of tedious steps, and no one individual is likely to have the skills and the time to perform them all, at least not in time to meet a publication deadline. A well-illustrated work requires finding good visual material, selecting images that both illustrate the text and possess aesthetic appeal, matching the visual image with the text, and writing captions to capture the significance of the image to the text. In addition, the visual material must be transformed from various formats into a single format that can be replicated for the author, the illustrator, the designer, and the publisher, and which allows the production team precisely to match visual material with captions and text.

Working collaboratively a group of professionals from several agencies and organizations have produced a book they each believe the people of North Carolina will enjoy. Members of the group from the North Carolina Office of Archives and History include Steve Massengill, iconographic archivist with the Archives and Records Section (A&R), who retrieved images selected from card catalog entries and placed print orders; Alan Westmoreland, photo lab supervisor with A&R, who processed the print orders; Susan Trimble, editor with the Historical Publications Section (HPS), who scanned and converted images to digital format; Lisa Bailey, editor with HPS, who proofread the final version of the book; and Donna Kelly, section administrator of HPS, who proofread the manuscript at all stages and insured that all of the production activities remained on schedule. Heather Overton, public information officer for the Department of Agriculture and Consumer Services, shared her knowledge of the department and critiqued the manuscript. Elaine Kurtz, commercial space administrator with the North Carolina State Fair, produced images from the fair's files. Wesley Wyatt, manager of the North Carolina State Fair, supported the project from its inception in a variety of ways, from providing meeting space to reading early drafts of the manuscript. Teresa Smith Perrien, designer, and Cheryl McGraw, sales consultant with Jostens, Inc., accommodated requests for production schedule revisions and worked closely with us to help meet deadlines. Without the patience, diligence, and dedication to this project of each and all of these people, there would be no book, and for their collective effort, we are truly grateful.

Introduction

North Carolina Commissioner of Agriculture Jim Graham sat behind his desk on a cold, dreary, blustery December morning in the year 2000, his last full year of service to the people of North Carolina. In less than one month, Graham, a political legend in his own time, would voluntarily vacate the office he had held for more than thirty years. A somewhat rumpled navy-blue suit, beneath which he wore a dark blue pullover, enveloped his large frame, only slightly bowed by age. Between the fingers of his right hand he held his trademark cigar, which he used like a conductor's baton to orchestrate his conversation. I had come to interview him about his role in promoting the North Carolina State Fair, since the 1930s a division within the Department of Agriculture and one in which he took great pride. He waved me into his office with an expansive gesture and welcomed me warmly,

As commissioner of agriculture from 1964 to 2000, Jim Graham continually sought to improve the state fair. He recognized the fair as the primary vehicle to interpret the state's agricultural economy— past, present, and future—to an increasingly urban population. Here Graham poses with a sand sculpture of himself at the 1999 fair. Photo courtesy of NCDA&CS.

asking that I excuse the unusual clutter, the result of preparations to vacate his office. For a few moments he engaged me in the down-home small talk of two North Carolinians with a professional interest in one of the state's great institutions—a strategy designed, I suspected, to test whether my credentials as a Tar Heel were genuine and if my knowledge of the fair passed muster. I dropped the name of a friend of my father's, a tobacco farmer long active in the North Carolina Farm Bureau, who, like my father, was now in his eighties. Graham immediately recognized the name and astutely positioned me among the families of rural and small-town eastern Cumberland County.

With my credentials established, I moved into the interview, asking him if the story I had heard about his meeting his wife at the state fair were true.

"Yes, it is," he smiled, as if taking pleasure in both his memory and my knowledge of the event.

"I wonder if we could begin the interview by you telling me about that meeting?" I asked, and he was off and running.

"It was in 1941. I remember it well. My wife-to-be, I met her at the North Carolina State Fair, at the waterfall, a little wildlife exhibit there, waterfowl and so forth. She was teaching home economics in Rowan County, that being my home, and I had run into her, but I never had really met her. She brought a group of children, her students, down here to the state fair, and I made arrangements to be out there, and happened to see her. I hitchhiked a ride back on the school bus, and I infuriated the little fella that was sitting with her, that came down with her, 'cause I took his seat going back. His name was Frank White; I remember it well. Anyway, we met there, and that was the beginning of a romance and then my wedding and my wife of fifty-four years. So, a lot of things can happen at the state fair and will happen."

And so they have, to the delight of generations of North Carolinians who have made the state fair North Carolina's largest and most beloved continuing social event. For a century and a half the North Carolina State Fair has educated, entertained, and enthralled North Carolinians from Manteo to Murphy, from West Jefferson to Wilmington. The state fair began as an effort on the part of antebellum North Carolina's leading planters and agricultural reformers to convince the state's

During Graham's long tenure in office, the state fair changed dramatically. Graham oversaw the construction of most major exhibit halls, the modernization of the fairgrounds' infrastructure, and the initiation of year-round use of the fair's facilities. Here Graham speaks to a child at the 2000 fair, his last as commissioner of agriculture and consumer services. Photo courtesy of NCDA&CS.

The fair's original waterfall, constructed in 1940, quickly became a fairgrounds landmark and for three decades served as an important meeting place for parties arriving separately or reuniting before departing the fair. Photo courtesy of A&H.

farmers and planters of the value of the practice of scientific agriculture. Within less than a decade it became one of the state's most popular institutions. At present the North Carolina State Fair remains among the oldest, strongest, and most colorful threads in the fabric of memory that binds a people to their state, and in the memories that bind North Carolinians to one another. It is the place to which North Carolinians have traditionally gone, and continue to go, to see and celebrate one another, an annual celebration of the way in which North Carolinians live and have lived, an enduring part of the heritage of the state and its people.

MELTON A. McLAURIN

Founding the North Carolina State Fair

The Fair before the Civil War

The North Carolina State Fair, begun in 1853, provided an enjoyable holiday for thousands of North Carolinians. It quickly became the biggest social event of the year in the capital city and, more importantly, in the lives of many of the state's rural families. Members of the fair's sponsoring organization, the North Carolina State Agricultural Society, organized the previous year, had hoped that the excitement of a state fair would lure North Carolina farm families to Raleigh by the trainload, but entertainment was not the society's primary reason for establishing the fair. Instead, members of the society hoped that the fair would serve as an educational event and a showcase that would promote agricultural reform and industrial development. A revitalized agricultural economy and newly developing industries, the members believed, would in turn contribute to the economic awakening then occurring within North Carolina. The fair would prove successful beyond its founders' most imaginative expectations. For the remainder of the nineteenth century it was perhaps the most effective means of introducing North Carolina farmers to pure-bred livestock, the latest farm machinery, and improved farming methods (and the superior crops that resulted from their use) and, with less

The only known representation of the original fairgrounds, located east of the State Capitol, shows six exhibit halls in the racetrack infield. Photo courtesy of A&H.

The creation of the North Carolina State Fair was part of a larger state-wide economic reform movement and a direct response to antebellum North Carolina's national reputation for agricultural and economic backwardness—a reputation that was, regrettably, well deserved.

success, of promoting the products of the state's fledgling industries.

The creation of the North Carolina State Fair was part of a larger statewide economic reform movement and a direct response to antebellum North Carolina's national reputation for agricultural and economic backwardness—a reputation that was, regrettably, well deserved. The state's farmers in the 1850s depended upon farming practices surprisingly similar to those used almost two hundred years earlier. Farmers still relied upon a combination of folklore, superstition, and information gleaned from almanacs to determine when to plant, cultivate, and harvest. Few fertilized their crops, and those who did knew little of the scientific application of fertilizers.

Prior to the Civil War, the vast majority of North Carolina farmers knew little about scientific agriculture, relying instead upon tradition, folklore, and, perhaps, an almanac to produce their crops. Cover of *Gale's North Carolina Almanac, 1830* courtesy of NCC.

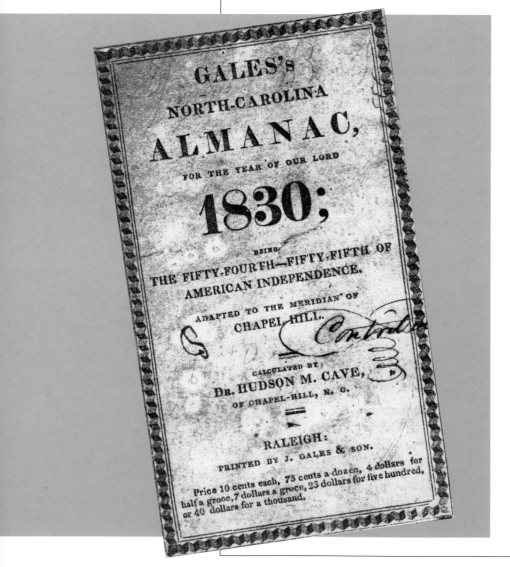

The average farmer was ignorant of how better methods of tilling the soil could improve the quality and quantity of his crop, and only a small minority of the state's farmers practiced such relatively simple and inexpensive measures to improve soil fertility as crop rotation and the adequate draining of fields. Despite higher levels of education, the state's plantation owners relied upon similar traditional agricultural practices and, in their dependence upon cash crops, were often more predisposed than were subsistence farmers to overcropping the land. Generations employing such crude farming practices had exhausted and eroded soils. Sadly, the average farmer and most planters did little to reclaim North Carolina's abused and increasingly unproductive lands.

Antebellum North Carolina farmers labored with badly outdated agricultural implements, for only a few of the state's more progressive farmers and planters possessed modern farm machinery. Workers on some of the state's largest farms still used the sickle to harvest grain. Knowledgeable critics described the average farmer's implements as "but the crudest, trashy productions of northern workshops."

Livestock received far less attention than did field crops, a fact underscored by the scruffy appearance of the state's farm animals. Most of the state's cattle were a mixture of several breeds, allowed to forage for themselves on an open range. Farmers also left sheep to fend for themselves and employed a system of "root, hog, or die" for swine herds that scoured forests and meadows for food. The common razorback or range hog formed the majority of the state's swineherds, which consisted of hundreds of thousands of animals, and only a few blooded swine were to be found.

North Carolina farmers and planters confronted several factors that had contributed to the state's retarded agricultural development. One of the most daunting was a serious transportation problem. North Carolina had no major port and few navigable rivers, and those streams that were navigable reached less than a hundred miles into the state's interior. For years the state government did little to help overcome this handicap, in part because the task of constructing an adequate transportation system was so challenging, in part because of a prevailing philosophy that private enterprise should develop it, and in part because those who held land, the basis of the state's wealth, resisted efforts to tax their holdings to finance better transportation. Few railroads existed prior to 1850, and those built later tended to parallel river systems rather than penetrate into new regions of the state's interior. Most of the state's roads were mere trails, usually filled with either mud or sand, which often transformed even short trips to market towns into frustrating misadventures. Such a transportation system severely hampered farmers' efforts to market crops and at times made marketing impossible.

In addition to its transportation problems, North Carolina lacked public schools until 1840, and for more than a decade thereafter the state's public school system existed on paper only. Not surprisingly, the majority of North Carolina's farmers possessed little formal education, and their ignorance hindered the state's agricultural progress. Lacking formal education, most farmers accepted without question the traditional agricultural methods used by their fathers. The state's warm climate, an abundance of game and fish, and a plenti-

and ignorance among the state's farmers resulted in generally low morale. Faced with the choice of reform, continued poverty, or migration, large numbers of North Carolina farmers and some planters abandoned the state beginning in the 1820s for the more fertile lands of the American Southwest, joining a mass migration to the frontier states of Mississippi, Louisiana, Arkansas, and Texas.

The plight of agriculture in antebellum North Carolina eventually forced the state's agricultural and political leaders to call for reform. Among the chief proponents of change were George W. Jeffreys, North Carolina's most prominent early agricultural reformer; Denison Olmsted, professor of chemistry at the University of North Carolina; and Paul C. Cameron, one of the state's leading planters and wealthiest men. Although these and other individuals recognized the state's agricultural problems and offered suggestions aimed at solving them, they lacked an organization to promote agricultural reform among the majority of the state's farmers.

There had been earlier efforts to promote agricultural reform, led by the planter elite, who relied heavily upon the production of the cash crops cotton and tobacco, both of which rapidly depleted the natural fertility of the soil. Planters established several county agricultural societies, especially in the east, beginning with the Edgecombe County Agricultural Society in 1810. At the societies' monthly meetings, prominent planters delivered speeches on agricultural subjects, and

The state's warm climate, an abundance of game and fish, and a plentiful supply of wild fruits and berries, conditions usually regarded as a blessing, proved a detriment to progressive agriculture. Some rural North Carolinians, finding that they could make a living without undue effort, were little inclined to become progressive farmers.

ful supply of wild fruits and berries, conditions usually regarded as a blessing, proved a detriment to progressive agriculture. Some rural North Carolinians, finding that they could make a decent living without undue effort, were little inclined to become progressive farmers.

In a state whose economy was almost entirely agricultural, such conditions were particularly damaging. Widespread poverty

members discussed mutual problems. Some county societies offered premiums for the superior production of certain crops and the best essays on agricultural topics, and during the 1820s several began to hold fairs in an attempt to encourage farmers to adopt agricultural reforms.

This flurry of activity at the county level led in 1819 to the formation of the North Carolina Agricultural Society, the prototype of the state fair's parent organization, the North Carolina State Agricultural Society, formed some thirty-five years later. The original state society offered cash premiums for essays written on agricultural subjects, such as the prevention and cure of certain livestock diseases, and for the best corn, wheat, rye, and cotton crops grown on two or more acres of reclaimed land. It also briefly considered

THE
CONSTITUTION
OF THE
AGRICULTURAL SOCIETY
OF
NORTH-CAROLINA;

ESTABLISHED AT RALEIGH, IN DECEMBER, 1818,

With a

LIST OF THE PREMIUMS

OFFERED BY IT'S

A List of its present Members;

AND

A Copy of certain Resolutions passed at the December meeting; 1319, with the Report thereon.

Published by order of the Society.

RALEIGH:
PRINTED BY J. GALES.

1819.

Efforts to bring about agricultural reform in the early nineteenth century led to the creation of the North Carolina Agricultural Society in 1819. The society considered sponsoring a state fair but rejected the idea and soon folded. Image courtesy of NCC.

sponsoring a state fair but rejected the idea. Never effectively organized, the state society became defunct within a few years after its establishment. County societies, composed of local men and organized on a more informal basis, struggled on, but their failure to attract the interest of "dirt" farmers and a general spirit of indifference led to their decline and disappearance during the 1830s.

Interest in agricultural reform reawakened during the 1840s, and county agricultural societies again appeared. The formality of earlier societies was replaced by informal discussion of agricultural problems, techniques, and the reform movement in general, and societies revived the practice of holding agricultural fairs

in a bid to attract the dirt farmer. In 1852, sensing the need to promote the growth of agricultural societies, the state legislature provided all county societies fifty dollars annually to support their reform efforts. In an effort to improve the ability of farmers to transport their crops to market, the legislature also extended an increasing supply of capital to various railroad and plank road companies. These improved transportation facilities helped the farmers to market their crops and generated further interest in improving the state's agriculture, while in 1852 the legislature began to address the problem of ignorance that plagued North Carolina's farmers by creating the state's first effective public school system.

As would be expected, the agricultural journals read by the state's farmers played an important role in contributing to the agricultural reform movement. Among the influential journals were Goldsboro's *North Carolina Farmer*, which began a four-year run in 1845, and the *Farmer's Journal*, begun in Bath in 1852 by Dr. John F. Tompkins, which lasted only two years. Edmund Ruffin's *Farmer's Register*, published in Virginia from 1823 until 1843, was probably the most popular journal read in the state, largely because of Ruffin's national fame as an agricultural reformer. Those journals and others included articles on all phases of farm life, urging farmers to adopt new and improved farming methods. Newspapers, too, devoted columns to agricultural reform.

County agricultural societies, journals, newspapers, and state government achieved some success in promoting agricultural reforms in the state, especially after 1845. Unfortunately, only small numbers of the state's wealthy, more progressive farmers adopted scientific farming methods. The great majority of the state's farmers remained unaffected, largely because the forces of agricultural reform generally failed to reach the average farmer. Although some of the state's smaller farmers expressed interest in agricultural reform, the more ardent advocates of new farming techniques remained a contingent of progressive planters, many of whom, as leading political figures, were also involved in several other aspects of the general reform movement sweeping the state. The need for some mechanism to introduce the reform spirit to the majority of the state's planters and farmers was obvious. The fairs of the county societies held the greatest promise of meeting that need.

Commercial fairs, the descendants of the old medieval trade fairs, were held in the South as early as 1723 and continued to be popular in North Carolina throughout the eighteenth century. The concept of the agricultural fair, however, originated with Elkanah Watson of Berkshire County, Massachusetts, who in 1809 displayed a pair of Merino sheep for the benefit of local farmers. Encouraged by the success of that exhibit, Watson and other farmers of the area in October 1810 staged a larger fair, generally considered the first organized agricultural fair held in the United States, and the following year organized the Berkshire County Agricultural Society. Largely because of Watson's efforts, by 1819 both Massachusetts and New York had granted state funds for premiums at the fairs of their county societies. The "Berkshire plan" soon reached North Carolina, and in 1821 the Rowan County Agricultural Society sponsored a fair described as "being novel in this part of the country." Other county societies held fairs in the 1820s, among them the Beaufort and Guilford societies. When the earlier societies failed in the 1830s, their fairs naturally suffered the same fate, but when county societies were revived in the 1840s, so were their fairs. The Mecklenburg society held an annual fair after 1842, and several other county societies soon did likewise. In addition, several regional fairs were established, and both county and regional fairs demonstrated an ability to attract the interest of North Carolina's farmers.

Many advocates of agricultural reform, especially editors of agricultural journals, felt that a state agricultural society and fair would provide a means through which agricultural reforms could be introduced to average farmers throughout the entire state. The editors of the agricultural journals were particularly strong supporters of this plan. Dr. John F. Tompkins, editor of the *Farmer's Journal* of Bath, made it his personal crusade to see that such a society and fair were established. He, more than any other individual, should be considered the father of the North Carolina State Fair.

Early in 1852 in the columns of his journal, Tompkins began a campaign to create a state agricultural society, praising such organizations in Pennsylvania and Maryland. He urged the farmers of North Carolina to establish agricultural societies in each county and to send, in July, delegates from each county society to Raleigh for the purpose of forming a state society. His proposal met with a thundering silence. In July 1852, smarting from the lack of response from farmers but determined and undaunted, Tompkins wrote: "They [farmers] have failed to notice the appeal, and we therefore plainly see that we have got the work to do our self." Because an extra session of the legislature was to meet that October, Tompkins merely altered his timetable and suggested that "the various County Societies appoint delegates to assemble at Raleigh on Monday the 18th of October next for the purpose of forming a State Agricultural Society," ending with the admonition that "every delegate who is appointed make it his business to attend the convention. . . ." In the September issue of his journal, Tompkins again implored the counties to form societies and elect as delegates to the convention men with "a deep interest in agriculture. We can assure them," he wrote, "that in giving their attention at this convention, they will be doing their country more service than by attending all the political mass meetings held in our state this fall."

Tompkins's determined efforts prevailed, and a State Agricultural Convention convened at Raleigh on Monday, October 18, at the

Commons Hall of the Capitol. Delegates represented Beaufort, Bertie, Brunswick, Buncombe, Carteret, Cumberland, Edgecombe, Greene, Guilford, Halifax, Haywood, Hertford, Johnston, Onslow, Pitt, Richmond, Rowan, Rutherford, Wake, and Wayne Counties. Wake had the largest delegation, Edgecombe the second largest. Most of the counties sent a single delegate, many of whom were leading planters of the county they represented.

The delegates elected Charles T. Hinton of Wake president pro tempore, but Tompkins remained its guiding spirit. They quickly adopted a motion by Dr. Tompkins that the president appoint five delegates to a committee to "prepare resolutions, and take necessary measures for the organization of the State Society of Agriculture." Hinton thereupon appointed as delegates Tompkins; John S. Dancy, a noted Edgecombe planter; A. J. Leach of Johnston; Lott W. Humphery, Onslow planter and politician; and Joseph G. B. Roulhac of Wake. The meeting adjourned until three o'clock that afternoon, at which time the delegates heard the report of the organization committee. That body recommended the formation of a state agricultural society, which was to have a president, four vice presidents, a recording secretary, a corresponding secretary, and a treasurer. It also recommended the founding of agricultural societies in every county and called for a committee of ten to be appointed to draw up bylaws and a constitution for the state society. The convention then elected a slate of officers for the newly formed state society, headed by John S. Dancy of Edgecombe as president. Following his election, Dancy appointed a committee of eleven to draw up the constitution and bylaws.

At three o'clock in the afternoon of Tuesday, October 19, Dr. Tompkins read those documents. They named the newly created organization the North Carolina State Agricultural Society, explained the duties of the various officers of the body and the rules of the organization, and the duties and obligations of its members and officers. The bylaws called for an annual state fair, to be held near Raleigh, and set forth detailed instructions to be followed in establishing the fair. The fair was to be partially financed by the society's annual five-dollar membership fee. The bylaws charged various committees, appointed by the president, with the work of organizing and promoting the fair. The fair's officers were to award premiums to "encourage a proper spirit of competition among the Planters, Farmers, and Mechanics of our country . . ." and to arrange for a speaker to deliver an address to the society at its annual meeting held during the fair week. Perhaps because of the reputation for rowdiness previous fairs had enjoyed, the bylaws empowered the president to appoint a chief marshal and five assistants, who were to "appear on horse back . . . to see that proper order is maintained [at the fair]."

In a bid to ensure the fair's establishment, the convention also created a five-member committee to present a memorial to the General Assembly requesting funds to enable the society to carry out its plans to hold a state fair in October 1853. The stature of the prominent men appointed to that committee perhaps best illustrates the relationship of the proposed fair to the general reform movement of the 1850s. The committee consisted of Thomas J. Lemay, agricultural journalist and noted Whig reformer; Richard H. Smith, Halifax County planter, state legislator, and Whig congressman; Calvin H. Wiley, state superintendent of public schools and a consistent advocate of reform; Lewis Thompson, Bertie County planter; Kenneth Rayner, progressive planter and Whig reform advocate; and, at the request of Wiley, Nicholas Washington Woodfin, planter and one of the state's leading political figures. After passing several other minor resolutions, the society adjourned until the following October.

Immediately upon its founding in 1852, the North Carolina State Agricultural Society appealed to the state legislature for funds with which to stage a state fair the following year. This 1852 memorial to the General Assembly requesting funds for the society was printed in *Senate and House Documents Printed for the General Assembly of N. Carolina at the Session of 1852.*

[HOUSE DOCUMENT, No. 16.]

A MEMORIAL.

To the Honorable, the General Assembly
of North Carolina :

Your Memorialists respectfully shew, that on Monday, the 18th day of October, 1852, there was formed in the City of Raleigh, a State Agricultural Society, composed of Delegates representing County Associations, and of citizens from different parts of the State, all interested in the great cause of Agriculture. the leading interest of North Carolina: That the Society was duly organized on a permanent basis, officers elected, and a constitution and by-laws adopted ; and that this Association, whose object is the advancement of the industrial interests of the Commonwealth, having made an auspicious beginning, it was deemed important to its continued existence and success, that it receive the countenance and support of the State : it was, therefore, resolved to memorialize your honorable body, on the justice and expediency of an appropriation from the State Treasury, to promote these objects ; and the undersigned were appointed a Committee to draft said Memorial.

In obedience to said resolution, your Memorialists respectfully solicit the attention of your honorable body, to this interesting subject, and earnestly request your favorable consideration of the same.

The five-man committee appointed to write the legislative memorial quickly accomplished its task, and the document was sent to the legislature in November. Short and to the point, the memorial pointed out that North Carolina lagged behind its sister states in agricultural and industrial development but noted "visible signs of an awakening among our people" that could be aided significantly by a properly endowed state agricultural society. It reminded the General Assembly that the majority of the state's citizens were farmers or mechanics and ended with the admonition that should the General Assembly refuse to aid the newly formed state agricultural society and its proposed fair, neither could hope to survive.

The General Assembly, responding to the political clout of the newly created state agricultural society's leadership, on December 27, 1852, formally incorporated the North Carolina State Agricultural Society. The act of incorporation authorized the society to hold property valued up to fifty thousand dollars and stipulated that the society's rules and bylaws were to continue in force until changed by that organization. So important did the idea of promoting a fair loom in the formation of the society, and in the legislature's decision to incorporate it, that the charter issued to the society required it to hold an annual fair to promote agricultural and industrial development in the state.

The legislature's decision to incorporate the North Carolina State Agricultural Society and provide it funds to aid in the establishment of a state fair buoyed the spirits of the state's agricultural reformers, creating among them a sense of enthusiasm and change. Full of hope for a better future, the state's agricultural leaders began to look forward to and plan for the first annual North Carolina State Fair, an event that would significantly boost the agricultural reform movement throughout the state and become the most popular and important social event in the lives of thousands of North Carolina's farmers and their families.

Above left: This meeting notice for an agricultural convention, called by Dr. John F. Tompkins, an agricultural reformer from Bath who published the *Farmer's Journal*, appeared in Raleigh's *Weekly North Carolina Standard* on October 13, 1852. The excerpt also includes an article on Gen. Winfield Scott, which reflected the sectional tensions of the time.

Above right: The October 27, 1852, edition of the *Semi-Weekly Raleigh Register* announced the formation of the North Carolina State Agricultural Society and listed its initial officers.

From Private Society to State Agency

Building the Structure

The Pre-Civil War Fair

The North Carolina State Agricultural Society wasted no time in preparing to stage a state fair in October 1853. Since the society had existed for less than a year and had only a handful of members, its ambitious plans seemed unreasonable; but its members were confident that the fair would open that fall. The society faced a formidable list of problems in bringing the fair to life so quickly, the most pressing of which was obtaining money—to purchase land on which to hold the fair, to construct buildings to house exhibits, and to provide accommodations for both exhibitors and visitors. Only the political, economic, and social standing of its members, many of them men of enormous influence within the state, made such an ambitious schedule possible. The society also had to create rules and regulations governing exhibits, establish entrance fees, and publish premium lists.

A lack of funds threatened not only the society's plans to establish the fair but also the very existence of the organization itself, inasmuch as the charter the legislature issued to the society failed to provide for the issuance of capital stock. Even worse, the legislature ignored the

Balloons released at the opening of the 1984 state fair. Photo courtesy of NCDA&CS.

society's appeal for funds, turning a deaf ear both to a memorial drafted by the society requesting funds and to a senate resolution to appropriate one thousand dollars to the organization.

With only thirty members by February 1853, each of whom paid five dollars to join, the society was essentially broke. That same month Dr. John F. Tompkins's *Farmer's Journal* appealed to farmers to join the society and implored the people of Raleigh to support the fair, pointing out that a fair would bring them financial benefits. The appeal met with little success, but Raleigh residents saw potential benefits of a fair, and members of the society persuaded the Wake County commissioners to provide funds for the proposed fair. The commissioners agreed to pay one-half of the

amount needed to obtain land and buildings for the fair, provided that the total cost did not exceed five thousand dollars. The commissioners required the society to match that amount with subscriptions and donations from members and other interested parties. The city of Raleigh matched the county's generosity, donating a tract of land to be used as the fairgrounds. The society did its part by securing loans from friends and members, who agreed to accept bonds issued by the society.

The society accepted from the city of Raleigh a sixteen-acre tract lying between Hargett and Davie Streets on the eastern edge of the city within a mile of the Capitol. Using funds obtained through its bond issue and from Wake County, the society constructed the fair's original buildings during the summer of 1853. Throughout the antebellum period, Floral Hall, the main exhibition building, one hundred by fifty feet in size, housed exhibits of household manufactures, floral arrangements, fruits, fancy needlework, pantry goods, and other miscellaneous goods, many of which were produced by the women of the state, who were also granted control over the mounting of exhibits in the building. The smaller Farmer's and Mechanic's Hall, at seventy-five by thirty feet, was used to display machinery, field crops, and agricultural implements. Besides the two main buildings, the society constructed near Floral Hall a "refreshment room" for ladies and sunk wells to supply fairgoers with water.

As it prepared for the fair's grand opening, the society worked to encourage attendance. Its members obtained the cooperation of the state's railroad systems, which was essential if the fair were to attract visitors from beyond the Raleigh area. The Wilmington and Weldon Railroad offered visitors a special fare of one-half the regular rate and agreed to carry all goods exhibited at the fair free of charge. To encourage exhibitors, the company ran a special train to pick up exhibits on the Saturday before the fair. The Seaboard and Roanoke Railroad likewise carried exhibits free of charge and offered special rates to fairgoers—and even reimbursed riverboats for the cost of transporting goods from the Plymouth area up the Roanoke River to railway stations. For the remainder of the nineteenth century, the state's railroads cooperated with the Agricultural Society in promoting the fair

Reduced Fares to Raleigh, N. C.

Account N. C. STATE FAIR, OCTOBER 14-19, 1929
VIA
SOUTHERN RAILWAY

FROM	One Day Only Thu., Oct. 17	Every Day 2-Day Limit	Every Day 6-Day Limit	Fair Rates Longer Limit
Salisbury	$3.25	$6.30	$7.10	$7.08
Spencer	3.25	6.20	6.95	6.93
Lexington	3.00	5.50	6.20	6.17
Thomasville	2.75	5.00	5.60	5.58
High Point	2.75	4.65	5.25	5.22
Winston-Salem	3.00	5.30	5.95	5.93
Kernersville	2.75	4.75	5.35	5.34
Guilford College	2.75	4.25	4.75	4.74
Greensboro	2.50	3.95	4.40	4.40
Gibsonville	2.50	3.20	3.60	3.60
Elon College	2.25	3.10	3.50	3.50
Burlington	2.00	2.90	3.25	3.24
Graham	2.00	2.80	3.15	3.15
Haw River	1.75	2.70	3.05	3.02
Mebane	1.50	2.40	2.70	2.69
Hillsboro	1.25	1.95	2.20	2.19
Chapel Hill	1.50	2.20	2.45	2.43
University	1.00	1.70	1.90	1.89
Durham	.75	1.30	1.45	1.43
Morrisville	.75	.65	.70	1.00
Cary	.50	.50	.50	1.00
Goldsboro	1.25	2.30	2.60	2.60
Princeton	1.00	1.80	2.00	1.98
Pine Level	1.00	1.50	1.70	1.68
Selma	.75	1.35	1.55	1.52
Wilsons Mills	.60	1.10	1.25	1.23
Clayton	.50	.75	.85	1.00
Charlotte		8.00	9.00	9.39
Concord				8.61
Kannapolis				7.91
Asheville				14.70
Morganton				11.36
Hickory				10.20
Newton				9.69
Statesville				8.46

Tickets at special fares in fourth column authorized account State Fair will be on sale at all stations in North Carolina and certain points in Virginia.

Dates of sale and final limit will be suited to dates of Fair.

by providing free transportation or reduced freight rates for fair exhibits and special fares for North Carolinians headed to Raleigh during fair week.

With the fairgrounds acquired, buildings constructed, some money in the treasury, and inexpensive transportation arranged for fairgoers and exhibitors, the society officially opened the gates to the first-ever North Carolina State Fair at 12:00 P.M. on Tuesday, October 18, 1853. Mindful of the need for funds to meet fair expenses, the society set entrance fees at 25 cents per person, $1.00 per carriage, and 50 cents per buggy. It also required all exhibitors to become members if they wished to contend for premiums, simultaneously creating a new revenue stream and increasing its membership. The *North Carolina Standard*, a Raleigh newspaper, estimated that more than 800 items were placed on exhibit, although the society's official premium list included only 583 entries. Winning exhibits in the various categories were awarded a grand total of $524 in prizes.

To the delight of the society's members, the press hailed the initial fair as a spectacular success and called for its continuation. The *Raleigh Register*, in an editorial that must have seemed a bit optimistic to society members aware of the organization's precarious finances, boasted that the reception given the first fair guaranteed the success of future fairs. The *North Carolina Standard* editorialized that the fair revealed the state's vast resources and demonstrated that North Carolina had begun to move to a higher rank among its sister states. A Wilmington paper, while praising the fair, noted that it should have begun twenty years earlier and proposed that the person responsible for its establishment be called a great benefactor of the state.

At its first annual meeting, which coincided with the 1853 fair and continued to do so throughout the society's existence, society leaders began planning the following year's fair. President John S. Dancy appointed a committee to ensure that the fairgrounds would be put in readiness for the 1854 fair.

Members elected Richard H. Smith as president for 1854.

Despite the ever-present lack of financial aid from the state, the society continued its preparations for the 1854 fair, adding more stalls for livestock exhibits and an amphitheater. With that construction, the fair's facilities were completed and remained essentially unchanged for the entire period during which the society held the fair on the original grounds. Desperately short of funds, society leaders somehow managed to raise enough from loans and membership dues to stage the 1854 fair. The second edition of the fair offered a slightly larger premium list, featured more exhibits, and drew larger crowds than did the first. Once again, both the society and the press acclaimed it an unqualified success.

In what was to prove its most significant action since establishing the fair in 1853, the society at its 1854 meeting elected Thomas Ruffin president. Ruffin, a lawyer, judge, chief justice of the North Carolina Supreme Court, and gentleman farmer, was one of the leading figures of antebellum North Carolina. He was a man of great ability and integrity and one of the state's most respected citizens. Ruffin remained president of the society until 1859 and during his presidency transformed the fledgling fair into a North Carolina institution supported by people of all walks of life and from throughout the state. When he began his term as president, however, the society's shaky financial situation threatened the fair's continued existence. The treasurer's report for 1854

As president of the State Agricultural Society for most of the pre-Civil War era, Thomas Ruffin of Hillsborough, a highly respected planter, jurist, and former chief justice of the North Carolina Supreme Court, provided the leadership required to establish the state fair. Photo courtesy of A&H.

revealed that the organization had in its treasury $4,886 received from gate receipts, dues, and donations. Unfortunately, that entire sum was required to cover the expenses of the 1854 fair. As a result, just as it had in 1852 and 1853, the society faced the new year with nothing in its treasury to finance preparations for and promotion of the 1855 fair.

Despite the fact that two successive highly successful fairs had been staged, in late 1854 it seemed that continuing financial problems would destroy both the society and its fair. In

expenses incurred in doing so had left it in debt. Other southern states, the memorial argued, had recognized the value of state societies and society-sponsored fairs and granted them aid. Could not North Carolina, the majority of whose citizens were farmers and mechanics, do likewise? Unless state aid were granted, the memorial warned, the society would fail, as would the fair it sponsored.

Responding to the society's lobbying efforts, the legislature on February 10, 1855, passed an act providing the organization an

Other southern states . . . had recognized the value of state societies and society-sponsored fairs and granted them aid. Could not North Carolina, the majority of whose citizens were farmers and mechanics, do likewise?

November 1854, Dr. Edward A. Crudup, chairman of the executive committee, asked Ruffin to call a special meeting of the society to see if state aid could be obtained. By December 2 the society was in debt in the amount of one thousand dollars, and Crudup recommended to Ruffin that the organization borrow money on the members' individual bonds, as had been done in 1853. Ruffin supported Crudup's suggestion, and, in part because of Ruffin's prestige, by the end of December Crudup had obtained the money required to ensure production of the 1855 fair.

Still, it was clear that the society needed state funds to guarantee that it and the fair would have a future. In January 1855, Ruffin called the special meeting Crudup had suggested. Although the society's membership remained small, its membership rolls continued to include the names of some of the state's most influential citizens, including a number of state legislators. The society urged its members who held legislative seats to encourage their fellow lawmakers to visit the meeting, which was held in Raleigh, in the hope that the visitors might obtain a better understanding of the society's goals and ambitions and be persuaded to grant the body some type of financial aid.

At the meeting, the society prepared and sent to the legislature another memorial requesting state funds. The memorial pointed out that the society had, through its own efforts, sponsored two successful fairs, but that

annual grant of fifteen hundred dollars (and requiring the society to match that amount) for premiums that would best encourage and promote the advancement of agriculture and industry in North Carolina. State financing was a vital achievement for the society, for it virtually guaranteed the sum of three thousand dollars with which to plan the coming fair. At that juncture, leaders of the organization sensed that support for the fair among Raleigh's citizens had diminished, creating another problem, since no other city was likely to support a failed effort and the location of the fair in Raleigh was crucial to the society's efforts to retain state funds. Such support was essential to the fair's survival. Only weeks before the 1855 fair opened, Crudup wrote Ruffin that the people of Raleigh evidenced a total apathy concerning the annual attraction. Moreover, he complained that they had done nothing to provide accommodations for fairgoers and had ignored his efforts to hold a mass meeting in support of the fair. More significantly, despite prodding from the society, Raleigh city commissioners had neglected to provide the body with a deed to the land that it had "donated" for the fair. At a special meeting in 1855, members of the society discussed this worrisome oversight. Dr. Crudup assured the members that the city of Raleigh had donated the land and that the city commissioners were prepared to issue a deed. Ruffin appointed a committee to secure the deed from the city, although a committee appointed for the same purpose at the 1854 fair had failed to do so. The commissioners refused the society's request and continued to do so until 1869, when the city finally granted the society a conditional deed. Crudup feared that if the attitude of the people of Raleigh did not change, the 1855 fair would be the last.

Nevertheless, the society, bolstered by the state appropriation, successfully staged the 1855 fair, which was well attended. By the fair's end, revenues from the state, gate receipts, and the sale of memberships left the fair with more than $4,500 in hand. Society membership had grown from 33 to 418 members, and membership fees now substantially aided the organization in meeting the fair's expenses. With dependable state aid and increased membership fees and gate revenues, the society in 1855 reduced its membership fee from five to two dollars in a bid to enroll and obtain the support of more of the state's "average" farmers. Although still far from affluent, the society had avoided financial disaster and could continue to stage its increasingly popular fair.

Under Thomas Ruffin's leadership the fair experienced rapid growth throughout the antebellum period. Each year saw larger crowds, more numerous exhibits, and increased exhibit and gate fees. In 1858, for example, a single day's gate fees totaled more than $900, and receipts for the entire week ran close to $3,000, some $300 above receipts for 1857. Each successive fair attracted larger numbers of North Carolinians, and in 1858 more than 8,000 people attended the week-long fair in an era during which Wake County had a total population of fewer than 25,000 persons. An increasing number of exhibits vied for the fairgoers' attention. In 1857 the fair displayed a total of 836 exhibits; by 1859 that number had increased to 1,344. In the same period, livestock exhibits grew from 203 to

252, exhibits in Planter's Hall increased from 144 to 246, and exhibits in Mechanic's Hall, in which industrial exhibits and craft exhibits were displayed, grew from 170 to 305. The society likewise experienced a steady growth in membership, enrolling 680 members by 1858.

Despite such obvious success, the society remained in financial distress and constantly struggled to find additional funds. At its 1857 meeting it offered a lifetime membership for a twenty-dollar fee in an effort to generate more revenue, although donations had increased that year, with eleven men giving fifty dollars each. Donations from individuals could not be relied upon as a revenue source, however, and members of the society felt that the state should increase its aid, contending that the Maryland and South Carolina agricultural societies received much more state aid than the General Assembly of North Carolina saw fit to give. In 1858 the society gave its executive committee a free hand to seek additional state funding by whatever methods it deemed necessary. The society's reliance upon receipts of a particular fair to cover that fair's expenses forced it to continually borrow to prepare for and promote the ensuing year's fair and kept the organization constantly in debt. Its inability to solve this problem meant that the fair, despite its growing popularity, remained in a precarious financial condition throughout the nineteenth century.

Although the fair continued to experience financial trouble, by 1859 it had become so popular and successful that other towns challenged Raleigh's status as host city. In 1859 Henry M. Pritchard of Mecklenburg County, a member of the state House of Representatives, attempted to amend a resolution in favor of the North Carolina State Agricultural Society by stipulating that the exposition be held in the same town no more than two consecutive years. Pritchard requested that Charlotte

The society's reliance upon receipts of a particular fair to cover that fair's expenses forced it to continually borrow to prepare for and promote the ensuing year's fair and kept the organization constantly in debt. Its inability to solve this problem meant that the fair, despite its growing popularity, remained in a precarious financial condition throughout the nineteenth century.

obtain the fair for the ensuing year if it would provide sufficient accommodations. Raleigh supporters rose to the challenge, and Pritchard's amendment failed by a vote of 53 to 32; nevertheless, the vote suggested considerable interest in seeing the fair held in a city other than Raleigh. At the society's 1859 meeting, Salisbury supporters moved that the fair be held in their city but withdrew the motion upon learning that it would be necessary for the legislature to amend the society's bylaws and constitution to make relocation possible. They succeeded, however, in having a committee appointed to determine if the legislature could make such changes to those documents.

Yet despite this note of optimism, the fair closed out the antebellum period in serious financial trouble. Terrible weather on the final day of the 1860 fair almost eliminated gate receipts, presenting newly elected society president William R. Holt with a deficit of nearly eight hundred dollars after paying all premiums and bills for upkeep of the fairgrounds. Like Ruffin, Holt was a member of the state's elite. Holt, the brother of Alamance County cotton textile industrialist and planter Edwin M. Holt, in 1860 owned more than 2,500 acres of land and fifty-nine slaves, making him one of the state's wealthiest men. He planned to hold a special executive committee meeting to attempt to solve the fair's financial crisis, but the coming of the Civil War soon overshadowed the society's concern with the problems of its fair, financial or otherwise.

The spectacular increase in the fair's popularity during the eight years of its antebellum existence, as well as its rapid physical expansion, occurred largely because the sponsoring North Carolina State Agricultural Society enjoyed such excellent leadership. In addition to Thomas Ruffin, such men as Weldon N. Edwards, Warren County planter, state

senator, and congressman; Paul C. Cameron, Ruffin's son-in-law and one of the state's leading planters reputed to be the richest man in North Carolina at that time; Kenneth Rayner, Bertie County planter, state legislator, and Whig congressman; and William R. Holt, brother of Edwin M. Holt, himself a prosperous planter—all noted and successful state citizens—gave their unqualified support, both political and economic, to the society and the fair. Ruffin, however, as president of the society for five of those eight years, was by far the leading figure in the establishment and success of the fair. His retirement from the bench in 1852 allowed him to devote full time to the management of his Alamance County plantation and his work in the society, especially after he was elected president in 1854. Merely by accepting the office of president, Ruffin lent considerable prestige to the society. He lost no opportunity to advance the society's interests, particularly in the matter of

Paul Cameron, Thomas Ruffin's son-in-law, was a politician and planter and the state's wealthiest citizen of the antebellum period, with landholdings in Wake, Granville, and Orange Counties. Cameron lent the fledgling state fair his prestige and financial support. Photo courtesy of A&H.

obtaining state funding, without which the fair surely would have failed.

Complementing Ruffin's effective leadership were the efforts of agricultural journalists within the society—men such as Dr. John F. Tompkins and Thomas J. Lemay, Raleigh editor, publisher, and champion of economic reform—who along with other agricultural journalists and newspaper editors enthusiastically supported the fair. Agricultural journals and newspapers served as the major means by which the society reached North Carolina's farmers throughout the year, not only during the month of October, and without their support the fair could not have prospered. Factors other than personal leadership added significantly to the growth of the antebellum fair. As previously mentioned, the fair benefited considerably from cooperation by the state's railroads, which agreed to transport fair exhibits free of charge and offered special rates to fairgoers.

A general reform spirit swept through North Carolina in the 1850s and created a climate of opinion that proved beneficial to the fair's successful establishment. Among a host of reforms carried out between 1848 and 1860 were the establishment of the State Hospital for the Insane, the adoption of free male suffrage, the completion of four major railroads, and the creation of the office of the state superintendent of schools. The spirit of reform spilled over into the state's agricultural life, as indicated by the appearance of five agricultural journals within this period. Although many of the state's farmers still clung to the outdated agricultural methods of their fathers, increasing numbers began to take an interest in agricultural reforms and the promise of scientific agriculture. These stirrings by North Carolina's farmers, a part of the general social and economic awakening of the state, created an audience for the fair that sustained its growth and development throughout the antebellum period.

War, Destruction, and Rebirth

When the Civil War engulfed the state, most of the North Carolina State Agricultural Society's members and officers rallied to the Confederate cause, certain that the conflict would soon be over and that victory would be theirs.

But visions of battlefield glory and a swift victory soon turned into the realization that the war would be a bloody, prolonged conflict—a realization ultimately to be transformed into a nightmare of death, destruction, and defeat. With its leaders and members fighting on the South's battlefields to defend chattel slavery and a way of life already bypassed by history, the society collapsed completely during the war years. As the war effort voraciously consumed what wealth and will the white populace of North Carolina and the South had amassed, the society and its fair were a luxury quickly abandoned, their rapid development and growing popularity overwhelmed by the increasingly desperate plight of a people at war.

The fair itself fell victim to the war, but its grounds and buildings served the Confederate cause. During the secession crisis, the Ellis Light Artillery, organized prior to the vote to secede, drilled at the fairgrounds. Immediately following North Carolina's secession from the Union in May 1861, the state transformed the fairgrounds into its first camp for the instruction of its hastily raised volunteers. Practically overnight some five thousand men from throughout the state poured into Raleigh to train to fight the armies of what they now regarded as their former country and present foe. The camp, initially under the command of Col. Daniel Harvey Hill of the North Carolina Military Institute in Charlotte, one of many Southern schools established in the two decades prior to the Civil War to train young men for military careers, was named Camp Ellis in honor of the sitting governor, John W. Ellis. That same month the state converted the fair's exhibit halls into a military hospital, known as the Fair Grounds

Daniel Harvey Hill, a native of South Carolina and a graduate of the United States Military Academy, chaired the mathematics department at Davidson College and in 1859 became the superintendent of Charlotte's North Carolina Military Institute. With the outbreak of the Civil War, Hill assumed command of Camp Ellis, a training facility located at the state fairgrounds, and eventually rose to the rank of major general in the Confederate army. Photo courtesy of A&H.

Hospital, and placed it under the direction of Dr. Edmund Burke Haywood, a Raleigh surgeon. The following August the Confederate government assumed control of the facility, and the Confederacy continued to utilize both the camp and the hospital until the war ended. The defunct State Agricultural Society paid a high price for the use of its fairgrounds and buildings by Confederate North Carolina, for victorious Union troops set fire to the structures during the closing weeks of the war, completely destroying some buildings and badly damaging others. At the end of the war, the society, like the Confederacy, no longer existed, and the fairgrounds lay abandoned, its buildings in ruins.

Resurrecting the Fair

The end of the Civil War did not result in the immediate reorganization of the Agricultural Society and the revival of its fair. Rather, the state's leaders, including many of the society's former officers and active members, devoted their time, effort, and funds to seeking solutions to a host of monumental economic, social, and political problems facing the state and its people. Deeply divided over issues of race and economics, North Carolinians struggled through the difficult Reconstruction years, which were marked by periodic outbreaks of violence and incessant political warfare. No effort was made to revive the North Carolina State Agricultural Society until four years after the close of the war.

Plans for reviving the fair surfaced in February 1869 as part of an effort to help unite a population that, after eight years

of war and Reconstruction, remained bitterly divided. Former members of the State Agricultural Society implored Thomas Ruffin to lead the effort to revive the body and its annual fair, for, as one put it, "the state now needs, more than ever, those pleasant and successful re-unions [of] before the war." Through the efforts of Ruffin and others, the society was reorganized in the spring of 1869, and the task of restoring the fair fell to the body's newly elected president, Kemp P. Battle. A member and officer of the society prior to the war, Battle was an extraordinarily talented lawyer, politician, scholar, and planter and a member of one of North Carolina's most prominent families. In later years, as president of the University of North Carolina, he became one of the state's most beloved public figures. To prepare for a fair in 1869, the society relied upon membership fees and the annual $1,500 subsidy from the state, which had been reinstated. Once again, the city of Raleigh came to the society's aid, finally selling it the old fairgrounds for one dollar (although placing some restrictions on the use of the property).

The society set about restoring the fairgrounds to its prewar condition, in some instances improving it. It enlarged the grounds, extended the racetrack to a length of one-half mile, and replaced and repaired burned buildings. The first postwar fair opened in October 1869 to an enthusiastic reception by both the public and the press. As a reward for Kemp Battle's leadership, the society twice reelected him president; Battle reluctantly accepted the third term when the society claimed to be unable to find another suitable candidate. Battle's reluctance resulted

Louisburg attorney **Kemp Plummer Battle,** a member of one of eastern North Carolina's most prominent families, served as president of the Agricultural Society from 1869 to 1872, presiding over the reestablishment of the fair after the Civil War. He was also instrumental in reopening the University of North Carolina following Reconstruction and served as president of the university from 1876 to 1891. Photo courtesy of A&H.

in large part from his success at reestablishing the fair, which within two years outgrew its restored physical plant. The buildings and grounds could no longer accommodate the attending crowds nor house the numerous exhibits. The society's appeals for increased state aid with which to improve facilities went unheeded. In addition, the fair faced competition from fairs sponsored by county or regional agricultural societies in Wilmington, Weldon, and other towns. As a result, the state fair experienced a sharp decline in popularity within four years of its seemingly successful reestablishment.

In 1872 the *Reconstructed Farmer* carried a devastating editorial attack on the fair, finding the grounds too small and in such disrepair as to be "disreputable to the State." It urged the citizens of Raleigh to improve the situation, suggesting that if they failed to do so the fair should cease to be called the state fair, be abandoned entirely, or be moved to another town such as Weldon, Goldsboro, Wilmington, or Charlotte. "As long as you have a Floral Hall better fitted for a shuck house than anything else, you need not expect an exhibition creditable to North Carolina," the editorial admonished, expressing sentiments echoed in other North Carolina newspapers and journals.

The Fair Finds a New Home

Faced with growing criticism, the leaders of the State Agricultural Society concluded that improvement to the fair's physical facilities must be undertaken if it was to survive. The task of revitalizing the state fair fell upon Thomas M. Holt, who in 1872 was elected president of the organization, a position he would hold for eleven years. Holt, a son of North Carolina's most famous antebellum textile manufacturer, Edwin M. Holt of Alamance County (and himself the owner of the Granite Factory, a cotton textile mill), was a member of one of the state's wealthiest and most powerful families. Like Kemp Battle, Holt was a versatile man, interested in industry, politics, and agriculture, although he was essentially an industrialist for whom agriculture was a secondary concern. He subsequently served as governor of North Carolina from 1891 to 1893.

Under Holt's leadership the society met its need for improved facilities by moving the fair to a new, larger site. In 1872 Raleigh resident Sallie E. Brown agreed to sell the society a twenty-two-acre tract from a parcel ideally situated on the Chapel Hill road, three-quarters of a mile west of the Capitol and adjacent to the rail lines of the North Carolina Railroad (presently an industrial and residential area on Hillsborough Street between Horne Street and Brooks Avenue and across from the North Carolina State University campus). In April 1873 the society purchased an adjoining, slightly larger, tract of land from Timothy Lee and his wife. It acquired its new fairgrounds with proceeds from the sale of its first site, made possible when the city of Raleigh agreed

Thomas M. Holt served as president of the Agricultural Society from 1872 to 1882. Holt, the son of Edwin M. Holt, a future governor, and North Carolina's leading textile manufacturer in the late nineteenth century, used his prestige, contacts, and wealth to promote the state fair, which moved to a new site in 1873. Photo courtesy of A&H.

to remove all restrictions from the society's original deed, an action prompted both by civic pride and by fear that the fair might move to another city. The society partitioned the old site into lots, which sold slowly, providing it with less cash than anticipated and requiring it to carry a heavy mortgage on its newly acquired property.

Construction of four buildings for offices —two at each of the principal entrances to the newly acquired site—began in the spring of 1873. The main building, a two-story octagonal exhibit hall, included a center section that became the new Floral Hall and two wings that became Planter's and Mechanic's Halls. The society also constructed a three-story grandstand, 300 by 44 feet in size, which contained offices as well as spectator seating, the second floor alone able to accommodate three thousand people. Additional buildings included a new Machinery Hall, a one-story building 200 by 44 feet, a two-story conservatory, a judges' stand, a portable speaker's stand, 200 stalls for horses and cattle, and

75 pens for sheep and swine. A one-half-mile racetrack completed the fair's new physical plant, at a cost of fifty thousand dollars. The new grounds provided the society a total of fifty-five acres, more than three times the acreage of the old fairgrounds, assuring room for growth. Such room would be needed, for the site remained the fair's home through 1925.

In 1873 the fair opened its new facilities to an enthusiastic reception, and from that year until the 1890s it continued to expand, the State Agricultural Society adding to its physical plant to respond to the fair's increasing popularity. In 1874 the society added one hundred stalls for cattle and horses and twenty-five pens for swine and sheep to house the growing number of livestock exhibits. It planted trees to beautify the grounds, dug additional wells to provide water for fair patrons and exhibits, and constructed a new road to the northeast entrance. A separate Department of Agriculture, Immigration, and Statistics was established in 1877. In 1879 the society constructed a building especially for the use of this new department, an event of enormous import for the fair's future, as the department soon became one of the fair's major exhibitors. The building originally provided the department with sufficient space in which to exhibit agricultural products from each of the state's counties and to present awards to winning exhibits in the classes for which it offered premiums. Leonidas L. Polk, among the state's leading advocates of agricultural reform, founder of the North Carolina Farmers' Alliance, editor of the *Progressive Farmer* (one of the South's most respected farm journals), and North Carolina's first commissioner of agriculture, had inaugurated the department's exhibits at the fair in 1877, the year in which the department was created.

The year 1879 also saw the creation of an institution not directly connected to the North Carolina State Agricultural Society or its fair but instead linked to both by the peculiar racial mores of the American South. In North Carolina the gradual triumph of the Democratic over the Republican Party in a struggle for state control began in 1870 with "Conservative" Democrats gaining control of the legislature. In the 1874 elections the Democrats strengthened their hold and in 1875 convened a constitutional convention. That gathering resulted in the adoption of a new state constitution that severely restricted the rights of North Carolina's African American citizens and ultimately led to their disfranchisement and the institution of racial segregation after 1900. Well aware that northern commitment to their cause was receding, some leaders within the state's various African American communities hoped to convince certain moderate and influential white political leaders that their race was "worthy" of inclusion in the political process and that African Americans would continue to make strides toward improving their economic position.

In 1879 Charles N. Hunter, a former slave who had become one of Raleigh's best-known and most politically active black educators, with his brother and some friends formed the North Carolina Industrial Association. As with the North Carolina State Agricultural Society, on which the Industrial Association was modeled, the major function of the new organization was to stage an annual fair. According to the Industrial Association's charter, its fair was intended to "stimulate industry, skill, economy, and thrift among the Negroes

Charles N. Hunter of Raleigh, a champion of racial justice and one of North Carolina's leading educators of the late nineteenth and early twentieth centuries, founded the Negro State Fair in 1879. The Negro State Fair was usually held on the state fair-grounds, and Hunter remained its guiding force until its demise after 1930, a consequence of the state legislature's failure to maintain funding. Photo courtesy of A&H.

of the state and to afford the highest possible degree of impetus to the general progress of the race." Hunter, like other prominent black leaders, held beliefs similar to those of Booker T. Washington, who founded Alabama's famed Tuskegee Institute in 1881 and from that position promulgated them to an eagerly receptive white audience, both in the South and throughout the nation. Hunter and other such "race leaders" believed that if blacks "proved themselves worthy," whites would "reward" them with improved political and economic opportunities. Whites, of course, welcomed the efforts of all African Americans to prove themselves worthy but refused to extend the expected "rewards" of equal educational and economic opportunities when they did so.

From its inception, the Negro State Fair showcased African American accomplishments, both to African American fairgoers and to invited white political leaders, who, during an era in which blacks continued to vote and participate actively in political campaigns, frequently attended. Like the State

Before 1900, white politicians supported the Negro Fair because it provided a convenient venue at which to appeal for African American votes. Leaders of the Agricultural Society and other whites supported it because it reduced, but did not eliminate, black attendance at the state fair. With the enactment of legislation mandating strict racial segregation by the 1901 General Assembly, this social function of the Negro State Fair became even more significant. Throughout the lifetime of the State Agricultural Society, the North Carolina Industrial Association continued to stage the annual Negro State Fair, usually during the week following the Agricultural Society's fair.

During the 1870s and early 1880s the increasing popularity of the state fair resulted in an increase in its financial assets. In 1877 the fair's management discontinued the old admission charge based on carriages and buggies and adopted a per-person admission price, set at 50 cents for adults and 25 cents for children, resulting in substantially increased gate receipts. In 1879, aided by a $2,000 subscription from the citizens of Raleigh, ever

The Negro State Fair was essentially a carbon copy of the Agricultural Society's fair, but on a reduced scale. It boasted its own agricultural, industrial, and educational exhibits, horse races, and carnival-like midway and attracted large crowds, primarily from the state's eastern counties.

Agricultural Society's fair, it also quickly became a much-appreciated source of entertainment for the state's African American populace, especially those in the vicinity of Raleigh and Durham. The North Carolina Industrial Association negotiated with the Agricultural Society for the use of its grounds and held most of its fairs there. It also managed to obtain an annual appropriation of $500 from the state legislature, at a time when the state was providing the Agricultural Society $2,500 annually to offset the cost of its fair. The Negro State Fair was essentially a carbon copy of the Agricultural Society's fair, but on a reduced scale. It boasted its own agricultural, industrial, and educational exhibits, horse races, and carnival-like midway and attracted large crowds, primarily from the state's eastern counties. It also served as a forum for some of the nation's most noted African American leaders, among them Frederick Douglass.

zealous in their efforts to keep the fair in their city, the society offered a premium list of $4,500 in cash. The highly successful fair of 1882 netted $8,000, and leaders of the society hoped that the large sum would allow it to reduce its debt by half.

The leaders' hopes were not fulfilled. Despite the increased receipts, the society found itself in desperate financial trouble by 1883, with a debt of more than $20,000. It had borrowed $10,000 from its president, Thomas M. Holt, and an equal sum from the North Carolina Insurance Company, and it owed money to other creditors as well. Although the society had been able to pay the interest on those debts each year, little headway had been made toward reducing the principal. The 1883 fair was only moderately successful, and some of the society's creditors began to demand their money. As a result, the organization assigned all of its cash to treasurer Leo D. Heart and

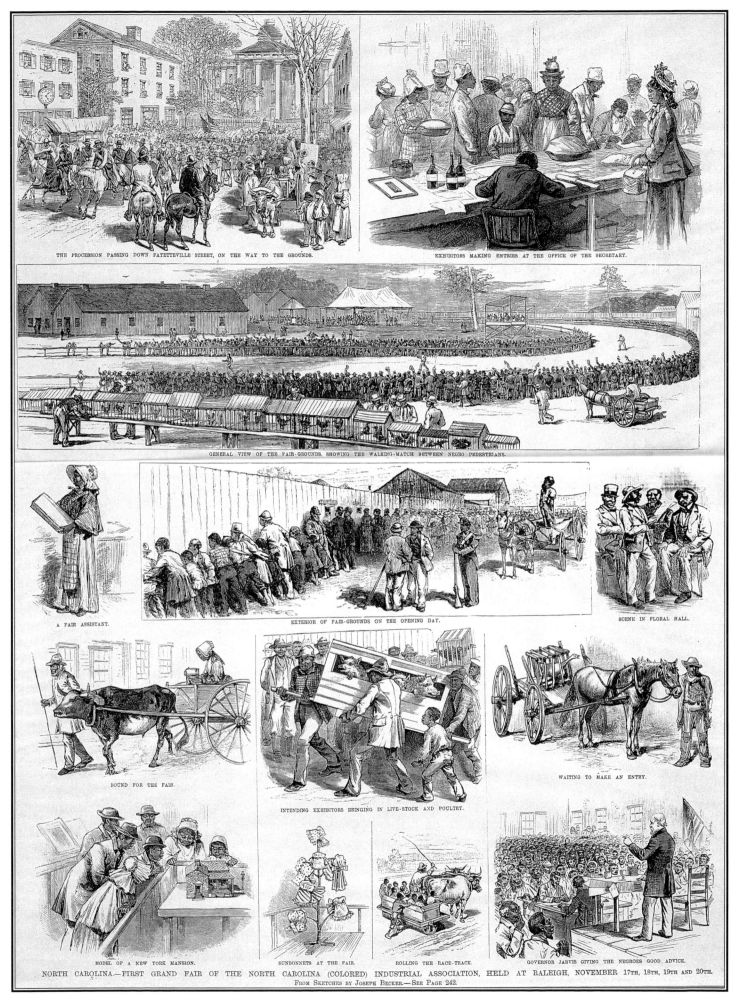

THE PROCESSION PASSING DOWN FAYETTEVILLE STREET, ON THE WAY TO THE GROUNDS.

EXHIBITORS MAKING ENTRIES AT THE OFFICE OF THE SECRETARY.

GENERAL VIEW OF THE FAIR-GROUNDS, SHOWING THE WALKING-MATCH BETWEEN NEGRO PEDESTRIANS.

A FAIR ASSISTANT.

EXTERIOR OF FAIR-GROUNDS ON THE OPENING DAY.

SCENE IN FLORAL HALL.

BOUND FOR THE FAIR.

INTENDING EXHIBITORS BRINGING IN LIVE-STOCK AND POULTRY.

WAITING TO MAKE AN ENTRY.

MODEL OF A NEW YORK MANSION.

SUNBONNETS AT THE FAIR.

ROLLING THE RACE-TRACK.

GOVERNOR JARVIS GIVING THE NEGROES GOOD ADVICE.

NORTH CAROLINA.—FIRST GRAND FAIR OF THE NORTH CAROLINA (COLORED) INDUSTRIAL ASSOCIATION, HELD AT RALEIGH, NOVEMBER 17TH, 18TH, 19TH AND 20TH.
FROM SKETCHES BY JOSEPH BECKER.—SEE PAGE 242.

Scenes from the first Negro State Fair, November 1879, from *Frank Leslie's Illustrated Newspaper*, December 6, 1879.

The North Carolina State Exposition. Opens Oct. 1st Closes Oct. 28th 1884. The main building covers more ground than any building ever erected in North Carolina. To be held at the city of Raleigh, N.C.

W.S. PRIMROSE, President.   H.E. FRIES, Secretary.   L.D.F. ARTT, Treasurer.

In 1884 North Carolina joined the ranks of nations and states that staged elaborate expositions to showcase their agricultural, industrial, and educational progress. The 1884 fair was incorporated into the State Exposition. Original poster for the exposition courtesy of A&H.

instructed him to pay premiums and current expenses from that sum and to give the remainder to A. Creech, Raleigh businessman; William G. Upchurch; Daniel G. Fowle, Raleigh attorney, state legislator, and governor from 1891 until his death that year; and Robert F. Hoke, Raleigh industrialist and longtime president of the North Carolina Railroad Company, each of whom was an assignee or trustee of a judgment against the society held by several banks and one individual.

The financial crisis caused by the modest success of the 1883 fair prompted the leading citizens of the state to propose a State Exposition for the following year. A delegation of northern visitors was expected during October 1884, and in an era in which southern states were mounting spectacular expositions in an effort to attract northern industries, state leaders feared that the fair could not sufficiently impress upon the northern capitalists the state's potential for agricultural and industrial growth. The Exposition, planned to run for the entire month of October, was to be

a giant display of the state's industrial and agricultural development specifically designed to impress upon the visitors and, coincidentally, the people of the state, the rapid growth of North Carolina's economy.

The State Exposition, planned and executed by private citizens, with residents of Raleigh most active in the planning, was held at the state fairgrounds. A huge, newly constructed Central Exhibition Hall housed the North Carolina Department of Agriculture's five-thousand-item North Carolina exhibit, which had been displayed at the 1883 Boston Exposition. Less significant exhibits appeared in other specially constructed smaller buildings. The total cost of the buildings erected for the 1884 Exposition was twelve thousand dollars, which was financed by private stockholders.

The 1884 Exposition was a success, attracting throngs of North Carolinians and visitors from other states. While no separate state fair was held that year, the Agricultural Society was legally obliged to offer premiums with the $1,500 obtained each year from the state, and one week of the Exposition was designated

fair week. The Exposition gave the fair a needed boost by arousing in North Carolinians a general interest in fairs and expositions. In the wake of the Exposition, the fair's growth quickened somewhat and continued to do so each year until the 1890s. By the late 1880s, it drew crowds of eight thousand to ten thousand people per day and featured thousands of exhibits from every region of the state.

With the exception of minor additions and repairs to existing buildings, the development of the fair's physical plant during the nineteenth century ended with two transactions involving the fair and two of the state's agricultural institutions. Those transactions contributed to Raleigh's successful bid for a proposed state agricultural college. In 1885 the state Department of Agriculture purchased from the Agricultural Society ten acres of land (at a price of fifty dollars per acre) to be used as an experimental farm by the department's Agricultural Experiment Station, which was created and funded in 1877 under the provisions of the federal Hatch Act and initially housed at the University of North Carolina. The society allowed the department to use, rent-free, twenty-five additional acres for the same purpose. In 1887 the society offered the North Carolina College of Agriculture and Mechanical Arts (later North Carolina State College) the use of that same land on similar terms. The stockholders who had financed the Exposition Building, located on the fairgrounds, donated the structure to the newly established college. Thus, by the end of the 1880s, while the Agricultural Society remained a private organization, its affairs were already linked to those of the state's two most important agricultural institutions—the Department of Agriculture and the North Carolina College of Agriculture and Mechanical Arts.

By 1890 the Agricultural Society, with justification, billed its state fair as the "Largest Fair in the South." The society now used modern advertising methods to promote the fair, including posters, handbills, free tickets, and all the free publicity it could arrange with an

This Wake County exhibit was typical of county exhibits entered at the 1884 State Exposition. Photo courtesy of A&H.

Additional buildings erected for the 1884 State Exposition that continued to serve the state fair included this dining hall. Photo courtesy of A&H.

accommodating press. Aiming its promotional materials at the typical North Carolinian, the society portrayed the fair as a place where the entire family could enjoy a day of good, clean fun. A whirlwind of social activities in the capital city during fair week contributed to the fair's ability to draw large crowds, as did the development of the midway, with its carnival attractions.

Yet despite the upswing in its popularity, the fair continued to be dogged by financial troubles. In 1895 the merchants and citizens of Raleigh donated to the society three thousand dollars to be used for the payment of back debts. The society spent this money, as well as the receipts from the 1895 fair, and required additional funds to complete repairs to several buildings, including the grandstands, in preparation for the 1896 fair.

In an effort to improve its financial situation, the society in 1895 elected as president Bennehan Cameron, gentleman farmer, leading social figure, and a member of one of the state's most prominent and wealthiest families. Cameron and his friends raised the funds required for the society to meet the expenses of the 1896 fair but only slightly reduced the organization's debt, and in January 1897 the society had on hand a balance of only $10.74. Despite the fund-raising efforts of Richard H. Battle (Kemp P. Battle's brother), elected president of the society in 1897, and Col. John S. Cunningham (prominent Person County planter and agricultural leader) elected president in 1898, the society was saddled with a bonded debt of twenty-six thousand dollars by 1899. Nevertheless, gate receipts that year rose by 30 per cent above those of the previous year, enabling the organization to make payments of interest on its bonds. Claude Baker Denson, for several years treasurer of the society, pointed out that the fair was improving financially and asked bondholders to wait for "the

Some of the state's most prominent men continued to lead the North Carolina State Agricultural Society in the late nineteenth and early twentieth centuries. Richard H. Battle, a Wadesboro attorney who, like his older brother Kemp, was devoted to the University of North Carolina, was elected president of the body in 1897. Photo courtesy of A&H.

inevitable advance in the price of bonds," which in 1899 sold at 25 cents on the dollar (indicating that original purchasers had made a contribution rather than an investment). Unquestionably, the society's failure to solve its financial problems was a detriment to the fair. To some degree, that failure reflected the ups and downs of the national economy, which experienced one of the nation's longest, most severe depressions during the 1890s.

On the other hand, the fair benefited from the enthusiasm for expositions and fairs that swept the South and the nation during the eighties and nineties. This was especially true of the Exposition of 1884, which helped renew interest in the state fair and led to direct state support of the fair through exhibits previously displayed at expositions in other states. The state's support was enhanced in 1877 when the state Department of Agriculture began its long, complex relationship with the fair.

The efforts of some of North Carolina's most prominent leaders, rather than financial support from the state, proved the most important factor in the ultimate success of the nineteenth-century postbellum fair. Thomas Holt, Kemp Battle, Bennehan Cameron, and others stepped forward to take the responsibility of guiding the fair through each serious crisis it faced, appealing to their wealthy and influential friends for the funds required to keep it and the State Agricultural Society afloat. They knew and were known by many of the leading industrialists and plantation owners of the state, and their contacts with such people drew many exhibits to the fair that otherwise would have been impossible to obtain. Although not primarily agriculturists, most of the fair's leaders—professional men, industrialists, entrepreneurs, and politicians—were all experienced with and capable of managing complex organizations. These men were, however, involved in agriculture to a certain degree and were sincerely interested in the fair and its ability to contribute to the well-being of the state. Because of their efforts, the state fair, despite financial trouble throughout the nineteenth century, experienced phenomenal increases in numbers of exhibits and attendance and by the turn of the twentieth century was firmly established and supported by citizens throughout the state. From a small concern with a $524 premium

list in 1853, the fair had become a complex organization requiring thousands of dollars to operate, bringing pleasure and instruction to thousands of people from every region of North Carolina.

Leadership from the Turn of the Twentieth Century through the First World War

From the turn of the twentieth century until after World War I, the fair's leadership, composed entirely of white men, continued to represent North Carolina's agricultural, industrial, and political elite. Positions of leadership within the Agricultural Society, as well as assigned responsibilities for various categories of fair exhibits, rotated among that prestigious group of North Carolinians. For example, in 1900 Charles McNamee, director of George Washington Vanderbilt's farms at Biltmore, served as society president. Serving as vice-presidents, each representing different regions of the state, were Kemp P. and Richard Battle, Julian S. Carr, Bennehan Cameron, and J. S. Cunningham of Person County. Joseph E. Pogue of Raleigh served for many years as secretary, and C. B. Denson remained as treasurer. The military titles carried by General Carr and Colonels Cunningham and Cameron reflected their status as revered Confederate officers and enhanced their appeal as leaders.

In 1905 Pogue and Denson remained in their positions, as did the Battle brothers, Carr, Cameron, and Cunningham. McNamee also served as a vice-president, along with Jacob A. Long, Person County attorney and politician, and W. R. Cox, Edgecombe County planter and lawyer. Ashley Horne, a Clayton planter, occupied the presidency. In 1910, in addition to its permanent vice-presidents, the society also elected representatives from the state's congressional districts. By 1915, as the United States sought to avoid being dragged into the conflict then raging in Europe, the society's leadership remained remarkably constant. Kemp Battle, Carr, Cameron, and Cunningham remained as vice-presidents, joined by J. H. Currie,

Right: While presidents of the Agricultural Society came and went, two highly respected Raleigh businessmen managed the organization's affairs, many of which concerned the staging of the state fair, during most of the late nineteenth and early twentieth centuries. Joseph E. Pogue served perennially as the society's secretary. Reproduced from the 1916 *Premium List*.

Below: Claude B. Denson served as treasurer of the Agricultural Society for much of the late nineteenth and early twentieth centuries. Reproduced from the 1916 *Premium List*.

Cumberland County planter, and E. L. Daughtridge, Nash County planter, merchant, and politician, both former society presidents. District vice-presidents who likewise had served in that capacity in 1910 included Graham County's Lynn Banks Holt of the famous textile family and Mecklenburg County's Sydenham B. Alexander, former president of the North Carolina Farmers' Alliance and one of the state's most respected agricultural spokesmen. Pogue and Denson, the two men most responsible for the smooth operation of the fair during this period, continued to serve as secretary and treasurer respectively.

As might be expected, the Agricultural Society's traditional leaders made few changes to the fair prior to World War I, relying instead upon the trusted formula developed during the late nineteenth century. The emphasis of the fair's exhibits continued to be on instructing the state's farmers in better agricultural practices. The premium list for the 1900 fair, for example, noted that the chief purpose of exhibits was to display the latest agricultural practices to the public, rather than to provide exhibitors with potential markets. Indeed, exhibitors were warned that "loud and noisy" efforts to sell their produce or livestock could result in exhibits being removed from the fairgrounds. Exhibits at the 1900 fair were divided into fourteen categories, including field and garden crops; a livestock category with divisions for horses, cattle, sheep, and swine; poultry; horticulture; pantry supplies; manufactures; and agricultural implements and machinery. Categories not directly related to the promotion of agriculture and industry included a general display; ladies' work; fine arts; educational, historical, and scientific displays; minerals; athletics; and gun contests. Those exhibit categories, with minor adjustments in groupings, remained in effect until well after the First World War.

America's entry into that conflict in April 1917 did not result in the cancellation of the fair that year. Rather, in addition to investing four thousand dollars in war bonds, the society took the opportunity to inform fairgoers of the "great necessity of the production and conservation of food and feeds" and to encourage the "raising of agricultural products and livestock." "Regarding all agricultural fairs as allies of our Government in its efforts to increase the raising of food supplies in the United States," the society pledged to the federal government "the hearty cooperation of our State Fair." In 1918, with America on a war footing, the federal government accepted the society's offer of cooperation, and that autumn the society sacrificed the fair to the war effort, as the United States Army had begun to construct a training camp on the fairgrounds.

In September 1918, the United States War Department established Camp Polk, a twenty-two-thousand-acre tank-warfare training camp, one mile west of the fairgrounds. The federal government had obtained the land for the camp through a lease with an option to purchase a large tract from some 120 farmers, which they accepted as a patriotic act in the time of war. This was a significant sacrifice for the farmers, inasmuch as their crops remained unharvested in the fields. The site encompassed the area approximately within the present boundaries of Hillsborough Street to the south, Glenwood Avenue to the north, the Raleigh Beltline to the east, and just west of the Raleigh-Durham International Airport to

the west. The present-day fairgrounds and the North Carolina State University School of Veterinary Medicine are located on portions of the site originally leased.

While the camp was under construction, however, the U.S. Army's Tank Corps used the fairgrounds, known as "temporary Camp Polk," to carry out its training for some five

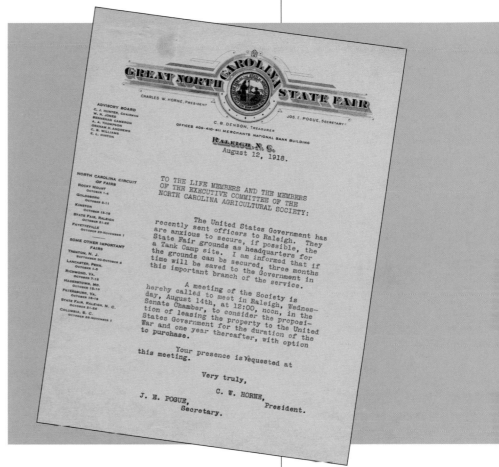

Above: During the First World War, the state fairgrounds served as a military base used to train troops in the use of tanks, which were from America's British and French allies. This letter of August 12, 1916, from leaders of the North Carolina State Agricultural Society urges life members of the organization to support that use. Image courtesy of Blankinship.

Left: A tank with its crew outside the fairgrounds Exhibit Hall in 1918. Photo courtesy of A&H.

During the latter part of 1918, thousands of troops were encamped in "temporary" tents erected on the fairgrounds while the U.S. Army rushed to construct Camp Polk, located approximately one and one-half miles to the west. Fortunately, the war ended in November, just as the construction at Camp Polk was getting under way.
Photo courtesy of A&H.

thousand troops. Workers hastily erected row upon row of temporary tents, with tar-papered wooden sides, each heated and electrically lighted and equipped with its own kitchen. The fair's grandstand became a garage for Ft-17 light tanks imported from France and Mark V heavy tanks imported from Great Britain, in which white soldiers trained. The army also constructed "overflow" camps approximately a mile from the fairgrounds at the rural community of Method. The segregated overflow camps accommodated white engineering troops and "colored stevedore troops." The War Camp Community Service, a voluntary agency, worked with local churches, the YMCA, the Knights of Columbus, and other Raleigh civic and fraternal orders to entertain the troops. Influenza, not Germans, proved to be the worst enemy of the Camp Polk troops as the epidemic of 1918 swept through their ranks, causing the base to be quarantined for much of the autumn. The quarantine was lifted on November 12, 1918, one day after the war ended in Europe, and by December only 310 white troops and 345 "colored" troops remained at Camp Polk.

Post-First World War Problems

After World War I the fair resumed in 1919 under the State Agricultural Society's traditional leadership and with the same structure as prewar fairs. In 1920 Gen. Julian S. Carr served as society president, Pogue and Denson remained secretary and treasurer respectively, and district vice-presidents included such society stalwarts as Bennehan Cameron, L. Banks Holt, and Sydneham Alexander, along with R. O. Everett of Raleigh, Leonard Tufts of Pinehurst, and C. W. Horne of Clayton. The fair entered the postwar years in reasonable financial condition, with assets far exceeding indebtedness and revenues from the 1919 fair approximating its expenses. Still, the society's wealthy membership continued the practice of lending the organization the funds required to prepare the ensuing year's fair.

This gentlemen's-club atmosphere, so effective in producing the fairs of the nineteenth and early twentieth centuries, was about to change drastically, leading to a

serious crisis that not only cost the society its fair but also threatened the organization's very survival. For reasons not entirely clear, the society, at its annual meeting during the 1920 fair, elected Edith Vanderbilt of Asheville as its first female president. Mrs. Vanderbilt, who succeeded Gen. Julian S. Carr, was the widow of George Washington Vanderbilt, of Biltmore Estate, which had been a major supporter of and exhibitor at the fair since before the turn of the twentieth century. She may have been selected as a means of expanding the circle from which the society recruited its members or perhaps because it was felt that she could obtain funds to improve the fairgrounds'

facilities, which had remained practically unchanged since their construction in 1873. Regardless of the reason she was selected, Mrs. Vanderbilt and the society's leadership (as Bennehan Cameron had done in the 1890s) bet that an increased emphasis on horses would attract additional members and fairgoers. In 1921, under Mrs. Vanderbilt's leadership, the fair staged a "Society horse show," with an emphasis on those breeds preferred by people wealthy enough to maintain horses as a hobby—hunters, jumpers, gaited saddled horses, harness horses, and ponies. Pleased with the horse show and in need of space for an expanding midway, the society

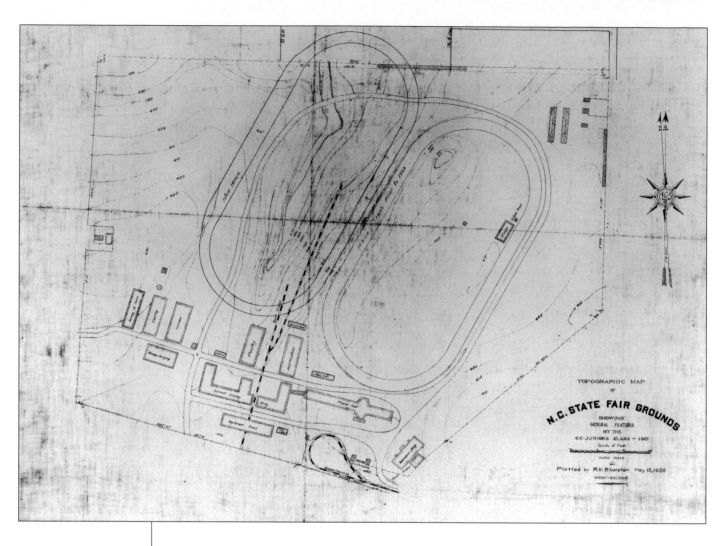

The expensive new racetrack constructed in 1922, shown here in a map of the fairgrounds, proved a bad investment. Unable to repay the funds borrowed for its construction, the Agricultural Society was forced to sell the fairgrounds and its facilities in 1926 in order to cover its debts, bringing to an end its history of sponsoring the state fair. Image courtesy of the Wake County Register of Deeds.

decided to construct a second racetrack, just to the west of the original track. Using its fairgrounds and facilities as collateral, it borrowed more than twenty thousand dollars on a balloon note that would come due in 1927. Given the fact that over the years only the wealth and generosity of the society's members had kept the fair afloat, this gambit represented a huge gamble. If the new racetrack did not attract enough new fairgoers to enhance gate revenues considerably, and do so relatively quickly, the society and its fair would face a potentially disastrous financial crisis.

Mrs. Vanderbilt's election to the presidency represented one of several fundamental changes in the society's leadership. By 1923, for example, Joseph Pogue, although he remained active in the society, no longer served as its secretary, having been replaced by Henry M. Landon, and a corporation, the Raleigh Savings Bank and Trust Company, served as society treasurer, replacing C. B. Denson. Moreover, the society had appointed a fair manager, E. V. Walborn of Raleigh. Josephus Daniels, the powerful editor of the Raleigh *News and Observer*, now served as a vice-

president, as did Pogue. The loss of Pogue as secretary and Denson as treasurer signaled the end of control of the fair by the society's old guard—and may have been prompted in part by a scathing 1922 auditor's report that blasted the society for the "careless and grossly negligent way in which both the money and the financial transactions were handled and recorded." More likely, however, the change in fair leadership simply reflected a generational change as the society's stalwarts grew old or died, to be replaced by younger men. Although somewhat younger, these men continued to represent the cream of North Carolina's elite and included such figures as Walter Clark, state supreme court justice, and O. Max Gardner, a rising young politician from Shelby whose political organization would come to dominate state-level politics until after the Second World War.

The exhibit categories, too, underwent major change. While such old standbys as livestock, field crops, manufactures, and agricultural implements and machinery remained, new categories such as vocational agricultural schools, boys' clubs, girls' clubs, and home economics were added. The additions reflected the expanding role of government agencies such as the North Carolina Department of Agriculture and North Carolina State College in planning, displaying, and judging exhibits. In 1920, for example, State College students mounted their first Student Agricultural Fair, which became a featured exhibit in subsequent fairs. By 1924 members of the North Carolina State College faculty served as superintendents of exhibits in the categories of horses, cattle, sheep, poultry, pets, eggs, field crops, vocational agricultural schools, and horticulture. Experts in the specific agricultural fields representing State College faculty, personnel from the Agricultural Experiment Station, and officials from the North Carolina Department of Agriculture now determined and judged exhibits, not leaders of the society.

The fair had become a complex organization, and the expenses associated with its operation continued an upward spiral. In 1924, for example, Southern Bell Telephone Company provided pay telephones for the fair; Western Union operated a communications center there; the Raleigh Chamber of Commerce dispensed information on the grounds; and the fair operated a post office substation, a press bureau, a children's nursery with rest rooms and nurses provided free of charge, a checkroom for wraps and packages, additional rest room facilities, and an emergency hospital—all services required by the thousands of people who now attended the fair each day it was in operation.

If the changes the society initiated had been designed to increase fair attendance and generate additional revenues to cover the costs of an increasingly complex fair, they failed miserably. By the end of 1922, the society had accumulated an $18,000 deficit, the result of the construction of the second racetrack and falling revenues. The following year, the society increased its liabilities with an operating deficit of nearly $6,000, despite having paid off notes issued to cover deficits from the 1922 and 1923 fairs. Only mortgages personally endorsed by Edith Vanderbilt and other society leaders covered the fair's expenses for the year. Facing a full-blown financial crisis with continuing operating deficits and mortgages on its facilities coming due within a year, the society in 1925 appealed to the state and the city of Raleigh for help.

With considerable prodding from powerful members of the State Agricultural Society and other agricultural spokesmen, the legislature responded with special legislation that called upon Gov. Angus McLean to appoint a committee to work with the society to stage the 1925 fair. McLean appointed O. Max Gardner to chair the committee, which consisted of 9 state representatives, 3 representatives chosen by the city of Raleigh, and 5 members from the Agricultural Society. The committee successfully staged the 1925 fair with funds raised by the sale of yet more bonds but did little to

O. Max Gardner, a graduate of North Carolina State College and a Shelby attorney, played a significant role in the transition of the state fair from a private organization to a governmental agency. Gardner, elected governor in 1928, continued his support of the fair, which came under the jurisdiction of the state Department of Agriculture in the same year.
Photo courtesy of A&H.

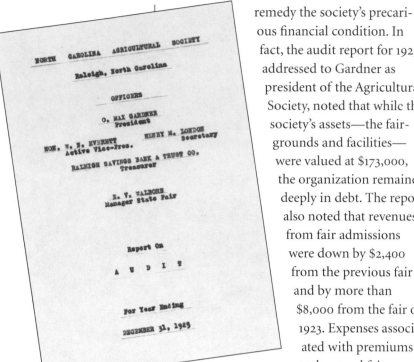

remedy the society's precarious financial condition. In fact, the audit report for 1925, addressed to Gardner as president of the Agricultural Society, noted that while the society's assets—the fairgrounds and facilities— were valued at $173,000, the organization remained deeply in debt. The report also noted that revenues from fair admissions were down by $2,400 from the previous fair and by more than $8,000 from the fair of 1923. Expenses associated with premiums and general fair operations, on the other hand, continued to increase. The auditors concluded that "It is evident, therefore, that something must be done to increase the Fair attendance to provide sufficient revenue to conduct it [the fair] properly."

The 1925 audit report proved a crushing blow for the North Carolina State Agricultural Society and thus threatened the continued existence of the state fair, for which the society had been responsible for more than seventy

years. Facing a mountain of debt and no prospects of increasing fair revenues, the Agricultural Society in November 1926 sold its fairgrounds and facilities to cover its debts. That decision reflected the financial reality that a private organization, even with a wealthy and powerful membership, could not afford to operate what had become, for all intents and purposes, a state institution. Although the State Agricultural Society continued to exist, its inability to produce the state fair cost it its reason for being, and it quickly became essentially an organization on paper only. The virtual demise of the North Carolina State Agricultural Society left the fair with neither a home nor a sponsor, and consequently no fairs were held in 1926 or 1927. Clearly, unless the state was prepared to rescue the state fair, an important part of the fabric of life in North Carolina was doomed.

The Fair Becomes a State Agency

For the next two years, state agricultural leaders lobbied the legislature to revive the fair and make it an official administrative unit within the state bureaucracy. This was an expensive proposal, since the society's sale of its fairgrounds and facilities meant that the state

As a young legislator in 1927, J. Melville Broughton, future governor and the son of a noted Raleigh businessman and politician, fought successfully to revive the state fair as an agency of state government under the control of the Department of Agriculture.
Photo courtesy of A&H.

the governor, served staggered four-year terms. Each of the state's congressional districts received a seat on the board, and three at-large members were appointed. The governor, the commissioner of agriculture, the director of the state Department of Conservation and Development, the president of North Carolina College of Agriculture and Mechanical Arts, and the mayor of Raleigh served as ex-officio board members. The legislation required that the fair be held on two hundred acres of land within five miles of Raleigh.

Fortunately, the state owned just such a site, readily available and clearly intended for the fair. During the First World War, the United States War Department had leased the land for Camp Polk, with an option to purchase. With the end of hostilities, the War Department decided not to make Camp Polk a permanent military base and revoked its lease on the property. In 1919, however, the state of North Carolina was able to exercise the purchase option that the federal government had held. The large tract that the state acquired included the grounds on which the North Carolina State University School of Veterinary Medicine is now located; the site of the former Polk Youth Camp, which was transferred to the State Prison Department in 1921 and became Polk Prison, later Polk Youth Center; and two hundred acres that became the current fairgrounds.

With a governing body and the fairgrounds obtained, all that remained was to find the money to construct new facilities and to operate the fair for the first year. Once again, officials of the city of Raleigh, concerned that the fair might depart the capital city and take with it the massive economic benefits it

would have to provide both for a new fair, bear expenses associated with their maintenance, and absorb the costs of planning and carrying out the annual fair. Those supporting the fair's revival found a champion in state senator J. Melville Broughton, member of a prominent Raleigh family and a future governor. Constituents unhappy about the loss of revenues brought about by the demise of the fair undoubtedly motivated in part Broughton's legislative efforts on behalf of reviving the annual event. During the 1927 legislative session, Broughton made an impassioned and, more importantly, a successful plea for the state to take over the fair, provide it with a new home, and underwrite it.

The General Assembly essentially passed legislation that attempted to placate all parties involved—state government, former members of the now nearly defunct North Carolina State Agricultural Society, and Raleigh city officials. The legislation placed the fair under the control of a board of directors that proved unwieldy both in terms of size and of interests represented. Board members, appointed by

An aerial view of the first buildings at the fairgrounds' current location in 1928. Photo courtesy of *N&O.*

generated, sought a solution, which was incorporated into the 1927 legislation. The use of the state's two hundred acres for the fairgrounds was conditioned upon the citizens of Raleigh raising $200,000 to "erect facilities on the grounds" and to help with operating expenses. Significantly, the legislature even provided a means for raising those funds. It empowered and authorized the city of Raleigh and the North Carolina State Agricultural Society to transfer to the state all funds received from the sale of the old fairgrounds and directed that those funds be used to finance construction of facilities at the new site.

The state fair's new board of directors selected W. S. Moye, a Rocky Mount businessman, to manage the revived fair, and Moye turned to the task of transforming a portion of the former Camp Polk site into the fairgrounds in time for the 1928 fair. Moye simultaneously oversaw the construction of the exhibit halls and livestock pavilions while arranging for exhibits and negotiating for midway and grandstand attractions. By opening day, "Hurry Up" Moye and his crews had constructed a grandstand and racetrack, a main exhibit hall, a women's building, pavilions for cattle, sheep, and swine, and

a poultry house, in addition to ensuring that the electrical, plumbing, and road systems on the grounds could meet the needs of fairgoers.

On Monday, October 22, Gov. and Mrs. Angus McLean and the red-coated North Carolina State College Marching Band led a parade of dignitaries, including the directors of each of the fair's departments and their wives, through the main gate, and the state fair was back in business. In his address to the opening day crowd of twenty thousand people, Governor McLean pointed out that the fair now belonged to the people of the state and could succeed only with their support and cooperation. Significantly, he addressed the fair's special role in showcasing to visitors "the experimental work of the State Department of Agriculture and the agricultural extension work of State College." He observed that the fair provided a forum for "the Department of Conservation and Development in connection with the conservation and exploitation of our natural and industrial resources. . . ." (To underscore the fact that the fair now belonged to the people of North Carolina, the state labeled the 1928 fair "the First State Fair" and continued that numbering scheme through 1938, reverting the following year to the practice of enumerating fairs from 1853.)

In linking the fair to the state's goals for economic development, McLean marked the transition of the fair from a private to a public institution. Under the auspices of the State Agricultural Society, the fair had been the state's most significant and well-attended social institution, designed primarily to promote scientific agricultural practices. As a state agency, the fair increasingly became a means of acquainting the people of North Carolina with the state government's plans for economic development. (McLean nonetheless acknowledged the continuing social significance of the annual autumn event, noting the fair's ongoing role in helping to bring together the people of the state's three distinct geographic regions.)

The most significant result of the state's acquiring the fair, however, was not the shift in emphasis that came with that acquisition but the very survival of the institution itself. If the state had not acquired the fair, it would never have survived. Even if, by some miracle, the North Carolina State Agricultural Society had found the funds to continue the fair throughout the 1920s, the depression that began in 1929 would have spelled disaster for both the Agricultural Society and the fair. The timing of the acquisition was perfect. The legislation making the fair a state agency passed in 1927, at the height of a period of strong economic expansion. The fair reopened under state auspices in 1928 on a new fairgrounds and with new facilities, exactly one year before the stock market crash that signaled the beginning of the Great Depression. Without question, under private sponsorship the Great Depression and the Second World War would have closed the state fair for at least a decade and a half. Whether it could have been revived by a private agency after so long a hiatus is highly doubtful.

The fate of the Negro State Fair reflects the difficulties a privately operated state fair would have encountered. The Negro State Fair was left homeless by the sale of the Agricultural Society's fairgrounds in 1926 and, like the state fair, was not held in 1926 and 1927. After the state fair was revived, however, African American leaders in the Raleigh area sought to resurrect the Negro State Fair, a task made more difficult because the legislature's annual five-hundred-dollar appropriation to

The Negro State Fair also resumed in 1928, but, unable to obtain state funds, it could not survive the depression. October 28, 1928 ad for the Negro State Fair from *N&O.*

the Negro State Fair had ceased with the demise of the North Carolina State Agricultural Society. Henry P. Cheatham, then president of the North Carolina Industrial Association, solicited a promise from Governor McLean to help the association stage its fair in 1928, using the new fairgrounds. McLean delivered on his promise, and the Negro State Fair was revived the week following the state fair. The four-day event opened on Tuesday, October 30, after workers had finished "getting exhibits into shape and refreshing numerous showings that were retained from the white fair." In an action that vividly portrayed the racist ideology of state government during that era, the state provided the Negro State Fair with second-class exhibits, decreeing that the state fair exhibits should remain in place. As a *News and Observer* reporter phrased it, "All of the State Department exhibits, which featured at the white fair, were left over." In 1929, however, the state, reflecting the racial prejudices of its white population, refused to allow the North Carolina Industrial Association to use the fairgrounds, and no Negro State Fair was held. Moreover, the state legislature refused to make its annual five-hundred-dollar appropriation to the association to support any fair it might hold elsewhere.

In one final effort to sustain an institution he had begun, Charles N. Hunter in 1930 per-suaded the North Carolina Industrial Association to empower him to stage the Negro State Fair using only the association's resources. Hunter, a veteran of the state's racial politics, convinced Gov. O. Max Gardner to direct the state Department of Agriculture to allow the Industrial Association the use of its fairgrounds, and the last Negro State Fair was held there under Hunter's management. Following the 1930 fair, Hunter tried in vain to convince the all-white legislature to reinstate a subsidy to support the Negro State Fair. With no appropriation, the association abandoned plans for a 1931 fair, and with that decision the Negro State Fair sustained a fatal blow. Charles N. Hunter died in September of the same year. Not only did African Americans lose one of the few statewide institutions available to them, but as a consequence of the state's segregationist laws they were also prohibited from attending the "white" state fair. For the next eighteen years, North Carolina's African American population remained a totally forgotten people at fair time.

World War II, however, created circumstances that would eventually lead to the inclusion of African Americans in the fair. The war heightened awareness and appreciation of the basic concepts of American democracy as the nation engaged in a death struggle with a totalitarian, racist foe. Even southern whites

Although African American fairgoers were clearly unwelcome at the state fair under the strictures of a segregated society, the fair was somewhat of an anomaly when it came to racial etiquette. Some African Americans continued to attend, through the fair's segregated exhibits. In this photo, African American children enjoy candy apples at the 1946 fair. Photo courtesy of *N&O*.

embraced the proposition that African Americans should be afforded a more "equal" place in society, but only within the confines of a "separate" status. While this view reflected racial prejudices long held by southern whites, it was also the settled law of the land.

With the end of the Second World War, black southerners, led by returning African American veterans, began to demand their full rights as American citizens. By 1948, civil rights had become a central, urgent issue in American politics, and white southerners doggedly defended segregation while taking steps to make things appear as if blacks were afforded "equal" treatment. In that year, state fair management, in what was undoubtedly felt to be a "racially progressive" step, created a separate department for "Negro" exhibits. The fair thus placed blacks in the humiliating position of having to embody "racial progress" at the segregated state fair. To enable blacks to demonstrate their economic and social progress to white fairgoers, the Department of Agriculture created Department Q, "Agricultural Extension Work with Negroes" as a category of African American exhibits and added it to the fair's premium list. Typical Department Q exhibits highlighted "the major activities of farm women, boys and girls enrolled in home demonstration and 4-H Clubs in North Carolina." The premium list recorded the fact that there were 17, 519 black farmwomen in home demonstration clubs and 38,422 boys and girls in 4-H Clubs. "Future progress and standards obtained by Negroes throughout North Carolina," the fair's management observed, "will, to a great extent, be influenced by the practical experience gained in Agricultural Extension Youth training programs."

In 1952, in a major reclassification of exhibits, fair management changed Department Q to Category E in Division One, General Exhibits, and titled the new category "Negro Home Demonstration and 4-H Club Exhibits." While the category title changed slightly over the years, it remained a fixture in

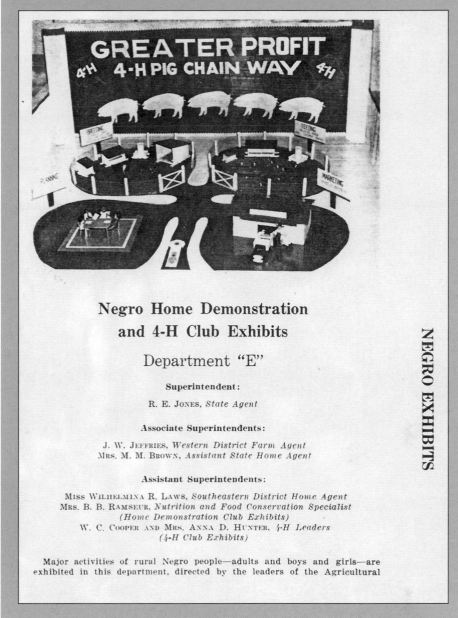

the fair's exhibits, usually under the direction of one of the state's segregated African American institutions. In 1953, for example, R. E. Jones, state Negro agent in charge of 107 farm and home demonstration agents in fifty North Carolina counties, was assigned responsibility for exhibits in Category E. He was assisted by leaders of the Negro Agricultural Extension Program, which was housed at North Carolina Agricultural and Technical College in Greensboro, the segregated (and unequal) equivalent of North Carolina State College. The exhibits reflected, according to the premium list of that year, "The major activities of rural Negro people, adults and boys and girls." The degrading exhibition of the activities of Negro Future Farmers, Negro Future Homemakers, Negro 4-H Club activities, Negro extension agents, and other African

In 1948, with African Americans in the South beginning to challenge racial segregation, fair management created separate exhibit categories for African American exhibitors. Department E was designated for Negro Home Demonstration and 4-H Club exhibits. Reproduced from the 1957 *Premium List.*

Americans struggling to survive in a segregated world, created for a largely ambivalent white audience at a segregated state fair, continued until 1965.

The state's assumption of responsibility for the state fair did not go smoothly. While the fair had a governing body, new grounds, and new facilities, lines of authority were not clear. Basically, the fair's board of directors acted as an ad hoc committee, outside the state's bureaucratic structure. In addition, money problems continued and dramatically worsened with the onset of the Great Depression in 1929, immediately following the fair of that year. To address both of those concerns, the state legislature in 1931 dismantled the elaborate structure created in 1927 and placed the fair directly under the auspices of the Board of Agriculture. Even that arrangement left the fair's managerial structure somewhat muddled, however, and did little to solve the fair's continuing financial problems.

Faced with the task of continuing to operate the state fair during the Great Depression, the Department of Agriculture repeatedly encouraged North Carolinians to support the fair. In 1931, for example, H. H. Brimley, director of the North Carolina State Museum, then a division within the department, urged North Carolinians to "Get the idea of attending the fair in your system, in your bones. . . . Fill up the gas-tank, crank up the good old bus that has served you so well, and climb aboard for a day at the new Old State Fair." In late September 1932, in an effort to bolster attendance, the Raleigh Chamber of Commerce sponsored a two-day bus motorcade that toured practically every town in eastern North Carolina. At each stop, participants urged people to attend the fair.

By the late 1920s, George Hamid of New York, a Lebanese immigrant and showman extraordinaire, controlled the booking of major circus and grandstand acts at venues along the East Coast. Hamid supplied grandstand acts for the North Carolina State Fair from the 1920s into the 1960s and actually leased and operated the fair from 1933 through 1936. Photo (ca. 1953) courtesy of NCDA&CS.

A Showman Runs the Fair: The George Hamid Years

Despite good attendance, the fair's money troubles continued. As the Great Depression continued with no relief in sight, the Board of Agriculture in 1933 considered closing the fair. The fair found an unlikely savior in the person of George Hamid, a New York showman extraordinaire who had supplied grandstand acts for

the fair since the early 1920s. Hamid operated by far the largest booking agency of carnival and fair performers in the East, booking fairs up and down the East Coast and in the Midwest and Canada, in addition to booking entertainment for Atlantic City, New Jersey's famed Steel Pier. In North Carolina, Hamid worked with Norman Y. Chambliss, a Rocky Mount businessman who operated the North Carolina Fair Operating Company, which booked fair acts for the Rowan and Sampson County fairs, as well as fairs at Rocky Mount, Williamston, Greensboro, and Pinehurst.

According to Chambliss, newly elected governor J. C. B. Ehringhaus summoned him to Raleigh in early 1933. Ehringhaus told Chambliss that because the fair was losing money, he felt that the state should "get out of the fair business." The governor asked Chambliss if he would be willing to lease the rights to operate the fair. Chambliss informed Ehringhaus that he lacked the capital to undertake such a venture but that he had contacts with someone who might be interested. He then contacted his friend and business associate in New York, George Hamid.

Hamid, believing that the closing of the North Carolina State Fair "would be a blow to the fair business all over the country," immediately drove to North Carolina accompanied by Max Linderman, a carnival operator. In Raleigh, Hamid spoke at a luncheon attended by the commissioner of agriculture and other dignitaries connected with the state fair. "I gave them the whole story," Hamid later wrote in his autobiography, "what was wrong with their fair and how to correct it, by driving out the gypsies and crooks. The result was more than I bargained for, they decided to hold the State Fair, provided that I'd run the whole thing myself." Hamid, in a tight spot financially because of falling revenues resulting from the depression, saw the offer as an opportunity to improve his financial situation. He accepted it and signed a lease to operate and manage the fair. The lease required him to promote the state fair, stage it, obtain exhibits and displays, and maintain the fairgrounds and facilities. Hamid agreed to pay the Department of Agriculture 15 percent of gross receipts, or at least $8,000, plus 50 percent of all net profits from the fair above $15,000. The Raleigh *News and Observer* roundly criticized the deal, denouncing it as a sellout to private enterprise and, worse yet, to a Yankee. To stem the criticism, Hamid contacted Josephus Daniels, the paper's publisher, and assured him of his plans to conduct a high-quality and successful operation. Daniels, according to Hamid, agreed to drop his complaints.

Thus, the first state fair of Franklin Delano Roosevelt's (FDR's) New Deal opened under the management of Hamid's partner, Norman Y. Chambliss, and Hamid paid a visit to Raleigh to personally supervise the event. The fair opened on a Tuesday, and Hamid had recovered his expenses by the close of the fair on the following Thursday. A Thursday-night fire threatened to send his anticipated Friday and Saturday profits up in smoke, however. The conflagration broke out in the main exhibit building and quickly spread to the archway connected to the industrial building, beyond which were the tents, shows, and rides of Max Linderman's carnival. Desperate to save his investment, Hamid corralled Linderman and together they led Ginger, one of Linderman's elephants, to the

flaming archway, threw ropes over the archway and attached them to Ginger. Ginger promptly pulled down the flaming structure, saving Linderman's carnival and Hamid's financial status. The publicity resulting from the fire and Ginger's heroics produced record crowds on Friday and Saturday, and Hamid recorded a profit that exceeded his most optimistic expectations. Profits from the 1933 fair placed Hamid once again on sound financial footing, and 1934 proved to be his best year since the onset of the Great Depression.

While the lease continued to prove profitable for Hamid, who operated the fair through the 1936 season, the Department of Agriculture clearly de-emphasized the state fair during his tenure. Coverage of the fair in official Department of Agriculture publications amounted only to exhortations to the public to attend and assurances that, although Hamid's organization had leased the rights to the fair, the department remained in control. Block ads appearing in the *Agricultural Review*, the department's newsletter, in 1933 urged farm families to "Bring your boys and girls, let them also explore the Capital City, it will help them with their schoolwork. . . . Take this Occasion to see RALEIGH with its State Departments, Colleges, Schools, and Various Other Centers of Interest." In its issue of September 25, 1934, the *Agricultural Review* carried only a banner announcing the dates of the fair, and in 1935 the publication merely noted that "The Fair now belongs to the State. Be Patriotic. Tell the Depression to 'go and stay put.'"

The Department of Agriculture's obvious lack of enthusiasm for the state fair under Commissioner William A. Graham, whom Gov. Cameron Morrison had appointed to that office in 1924 to succeed his deceased father, Maj. William A. Graham, did not sit well with some ambitious agricultural leaders, among them a young Alamance County dairy farmer named W. Kerr Scott. Scott was no ordinary farmer. A graduate of North Carolina

Top: Alamance County dairy farmer W. Kerr Scott, the son of a longtime member of the State Board of Agriculture and a graduate of North Carolina State College, was elected commissioner of agriculture in 1936. As promised, he persuaded the legislature to make the state fair a division within his department. This photo of Scott as commissioner of agriculture dates from 1939. Photo courtesy of A&H.

Bottom: Scott continued to champion the fair after his successful 1948 campaign for governor and following his election to the United States Senate in 1954. Here, as governor, the "Squire of Haw River" poses on his farm. Photo courtesy of A&H.

State College, he was also the son of a prominent Alamance County politician who had served for decades on the State Board of Agriculture. In 1936 Kerr Scott decided to challenge Graham for the position of commissioner of agriculture, and he seized upon the leasing of the state fair as a symbol of Graham's lack of dedication to the dirt farmers of North Carolina. The fact that Hamid, a New Yorker, operated the fair, lent considerable emotional support to Scott's campaign to return it to the control of North Carolinians and, of course, the state government department that he hoped to head.

Scott vowed to return the state fair to the people of North Carolina by making it a division within the Department of Agriculture and providing it with a permanent manager. Young, handsome, and a forceful speaker, Scott hammered away at Graham's record with the fervor of an old-time evangelist, his standard stump speech appealing unabashedly to state pride. "It is a disgrace that North Carolina must go to New York to secure men and finances to operate our fair," he told a Wake County audience. "We need a revival of the old-time agricultural spirit of our State Fair to displace that spirit of carnival and revelry with which it has been permeated for years." Scott's campaign, with the fair as a focal point, worked perfectly, and Graham went down in defeat.

The Modern State Fair: The J. Sibley Dorton Era

True to his word, once installed as North Carolina's new commissioner of agriculture in 1937, the young Scott successfully lobbied the legislature to make the state fair a division within the Department of Agriculture. Scott chose brilliantly when he selected the department's fair manager, for the man he chose would shape the destiny of the fair for the remainder of the twentieth century. J. Sibley "Doc" Dorton, a young man from Shelby, North Carolina, brought impressive credentials to the position, not the least of which was a close connection to the "Shelby Dynasty," the political machine of former governor O. Max Gardner and Gardner's brother-in-law, Clyde R. Hoey, who was elected to the governorship in 1936 (and to the United States Senate in 1944). But Dorton, a visionary and a born showman, also brought on-the-job experience to his new position, having managed the Cleveland County Fair since 1924. To the astonishment and delight of Scott and the Board of Agriculture, Dorton oversaw the 1937 fair and reported a net profit of more than $8,000, generated from fees from the carnival and grandstand show operators and increased gate revenues. Dorton actually increased premiums paid at previous fairs, expending $12,604. But the larger premiums offered increased public support of the fair, which resulted in doubling the number of exhibits from the previous year's fair. Profits would have been higher had not Dorton spent an additional $7,500 on improving the fairgrounds' electrical system, walkways to the grandstand, and exhibit areas in the industrial building. Such an achievement in the depths of the Great Depression underscores the success of Dorton's approach to return the fair to the people of the state and to make it a reflection of their accomplishments and aspirations. Until Dorton's death in 1961, his philosophy never wavered, and the expansion and profitability of the fair continued to reflect his remarkable administrative skills.

Having successfully launched the fair under the auspices of the Department of Agriculture, Dorton immediately began to campaign for the fair of his dreams, which was far from a week-long event held each October. During the 1938 fair, Dorton announced his plans to the Board of Agriculture. The fairgrounds and its facilities, he told the board, "can profitably be used throughout the year for permanent exhibits depicting the agriculture, educational, and industrial resources of North Carolina." What Dorton had in mind was a much grander type of attraction, one that would draw not only North Carolinians but

also tourists from throughout the nation to see displays of the state's natural history and all its resources. Aware that such a permanent exposition would be expensive and that the state lacked the resources to build it, he proposed that North Carolina apply for federal funds to finance his dream fair. Such a proposal would have seemed perfectly logical to Dorton, for he was well aware that Franklin Roosevelt's New Deal programs such as the Works Progress Administration, the Public Works Administration, and the Farm Security Administration had spent millions of dollars on cultural projects. He even credited Roosevelt with the idea of a permanent fair, noting that as governor of New York FDR

J. Sibley "Doc" Dorton, appointed the first manager of the state fair in 1937, was without question the father of the modern fair. Dorton, who had previously managed the Cleveland County Fair, foresaw the economic prosperity of the post-Second World War years and dreamed of a state fair that would serve the people of North Carolina throughout the year. "Doc" Dorton in his office at the state fair, ca. 1953. Photo courtesy of A&H.

The original waterfall, shown here under construction in an October 6, 1940, newspaper photo, was an immediate hit with fairgoers and became an instant landmark. Reproduced by permission of the *News & Observer* of Raleigh, North Carolina.

had observed that having the state fairgrounds sit empty and unused for fifty-one weeks of the year seemed wasteful.

The Second World War interrupted Dorton's ambitious plans, and he went off to serve his country as the North Carolina director of the War Manpower Commission, but not before he unveiled a towering waterfall, its cascading waters illuminated with multicolored lights, at the 1940 fair. The waterfall was an instant success with fairgoers and for decades was the gathering spot for families, exhausted from a day at the fair, ready to begin the journey home. The 1941 fair, organized around the theme "National Defense," would be the last one held until after the war years. Largely because of the rapid development of enormous military bases elsewhere in the state, especially Fort Bragg and Camp Lejeune, the fairgrounds was not used in direct support of the war effort, as had been done during the First World War. Rather, the fair rented its facilities to various firms for storage of tobacco purchased through the New Deal's Commodity Credit Corporations program, with up to ten million pounds of tobacco stored on the grounds at any one time.

With the war drawing to a conclusion, Dorton renewed his efforts to transform the fair from a six-day-long farm exhibits show to "a three million dollar State Exposition operating every week of the year, embracing all phases of North Carolina's governmental, agricultural, and industrial activity." Ever the shrewd showman and politician, Dorton pitched his vision to the 1945 state legislature as a means of "honoring veterans of two world wars" and succeeded in obtaining legislation allowing the fair to borrow $100,000 in bonds payable from fair receipts and to use the funds to develop plans for his building program.

Dorton dreamed big. At the heart of his planned fair of the future was a ten-thousand-seat coliseum, to which Dorton attached an amazing variety of bells and whistles, perhaps because he was a born showman but also, undoubtedly, to help sell his project to the state legislature, to which he turned for funds to build it. Dorton intended the coliseum to be used for conventions of agricultural, commercial, and industrial associations; for cattle, horse, and poultry shows; and, with an amphitheater to house all types of large events, to accommodate everything from auto shows to exhibits of farm machinery.

The coliseum would not only be functional; it would be beautiful, architecturally inspiring, and an announcement to the world of the greatness of North Carolina's people.

Dorton envisioned the coliseum as including a war memorial and being the
most beautiful building in the state,
futuristic in design, towering above all
other fair structures. The memorial
would house exhibits of war and war-
supportive products from every
county, as well as portraits of all
North Carolinians killed in the two
world wars. In addition, one hundred
massive pylons would have engraved
upon them the names of persons from
each of North Carolina's counties who
served in either war. "Reveille" and
"Retreat" would ring out each day
from bells atop a carillon tower as veterans raised and lowered the national
and state flags.

An early model of
"Doc" Dorton's pro-
posed "Fair of Tomor-
row," with a stadium
in the right-hand
background. Photo
courtesy of NCDA&CS.

Dorton promoted his fantastical concept of
the fair of the future even as he prepared to reopen the fairgrounds for the 1946 fair. In May
of that year he met with Gov. R. Gregg Cherry,
members of the State Board of Agriculture,
and agricultural leaders from throughout the
state to tout his vision, obviously with the
approval of Commissioner Scott. During the
meeting, U. B. Blalock of Wadesboro, chairman of the House Public Buildings Committee, pledged to Dorton that he would
recommend a state fair building program

Opening day at the 1946 state fair, the first after the Second World War closed the fair for four years.
Photo courtesy of N&O.

to the 1946 legislature. The pledge proved to be the first of several unsuccessful efforts to obtain the funding necessary for the fulfillment of Dorton's dream.

Meanwhile, the 1946 fair reopened under Dorton's leadership after its exhibit halls had received a face-lift. Commissioner Scott observed that the war had torn the fair asunder. But now, in Scott's words, "it is time for us to become acquainted with each other again. . . . At the State Fair we can meet together and talk together as in the peaceful days before 1941." For the next five years, while Dorton continued to lobby for this fair of the future, he directed the staging of the annual exposition in the facilities that he constantly sought to upgrade, while each year increasing the amount of money spent on premiums to entice better and more numerous exhibits. The 1947 fair, for example, boasted a new water system, fed by a water main from the Raleigh city system. Previously the fair had relied upon three nearby wells for its water supply.

Finally the political stars aligned, and Dorton could begin to create his long-held vision of a new state fair. In 1948 W. Kerr Scott successfully ran for governor and used the position to advance Dorton's plans for the fair. In 1950 Scott, Dorton, and their supporters carried the day, and the legislature, undoubtedly mindful that the fair would celebrate its

centennial in just three years, appropriated two million dollars to begin construction of a somewhat scaled-back version of Dorton's 1945 plan. At the 1950 fair, Dorton unveiled a model of the fairgrounds of the future, which, according to the Department of Agriculture, "disclosed ideas of architectural design so utterly new and daring in conception that they brought expressions of pleased amazement" from many who saw it. The model portrayed a master plan for the fairgrounds' entire 228 acres, including a 9,500-seat coliseum, a refurbished grandstand with seating for 10,000, new exhibit halls and cattle barns, a beautifully landscaped racetrack area, and a 100,000-seat sports stadium. Late in the year, the department let bids to begin construction of those aspects of the plan to be financed by the two-million-dollar legislative appropriation. Those projects included the coliseum; remodeling the grandstand; a youth center to house young fair exhibitors; conversion of the State Highway Commission shops adjacent to the original fair site to a livestock pavilion for up to 800 cattle, sheep, and swine; landscaping the racetrack with a series of pools; and grading of a new midway and parking areas. Proposed "for future development on a self-liquidating basis" was the construction of the sports stadium, a series of industrial exhibit buildings, a farm machinery pavilion, and a

Drawing of an improved fairgrounds with a visually stunning arena, which Dorton saw as the centerpiece for his modern fair, and a large stadium on the left. The shape of the arena, yet to be built, was already emerging as the symbol of the post–World War II fair. Drawing courtesy of NCDA&CS.

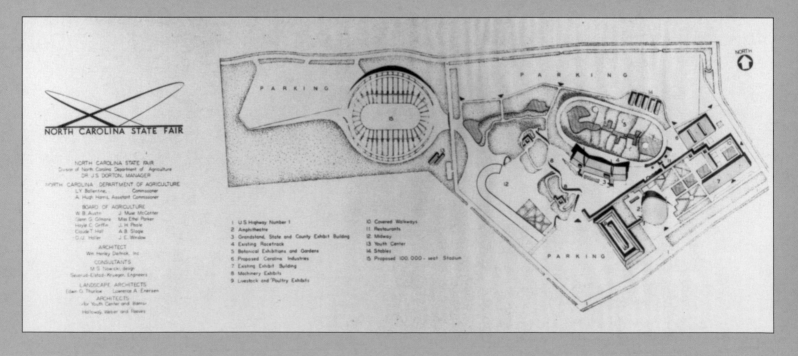

"deluxe café-club house overlooking the race track"—at a projected cost of approximately six million dollars.

As would be expected, North Carolina State College personnel and Raleigh businessmen figured prominently in the construction of the new fair. The project's centerpiece was the 9,500-seat coliseum, an elliptical structure with a roof supported by steel cables suspended from two intersecting parabolic arches. Matthew Nowicki, a Polish émigré who had trained at both the Chicago Art Institute and Poland's Polytechnic of Warsaw and was the acting chair of State College's Department of Architecture, designed Dorton's futuristic arena. Nowicki's radical design reflected concepts then fashionable among Europe's most admired architects, who were busily creating dazzling structures for the post-World War II world, including capital cities in the newly emerging nations of the old colonial empires. Nowicki had received awards for his designs at both the Paris Exposition of 1937 and New York's World's Fair of 1939 and after the war had worked as a consulting architect on the United Nations Building in New York and on the plans for Chandagarh, the new capital city of India's Punjab province.

Unfortunately, Nowicki never saw his futuristic building, for in August 1950 he was killed in an airplane crash in Egypt while returning from the Punjab. His arena plan was carried out by William Henley Dietrick, Inc., a Raleigh architectural firm headed by Dietrick, a close friend of Nowicki, in consultation with Stanislava Nowicki, Matthew's widow, herself a noted architect. The William Muirhead Construction Company of Durham served as general contractor, and Raleigh firms provided the air-distribution, plumbing, and electrical systems for the novel structure.

Above: Matthew Nowicki, a Polish-born architect trained in Europe and the head of the Department of Architecture at North Carolina State College, designed the futuristic arena that Dorton envisioned. Photo (ca. 1949) courtesy of A&H.

Left: Nowicki died in a 1950 airplane crash, but his widow, Stanislava (also a trained architect), shown here with her husband, helped to oversee the completion of the arena. Photo (ca. 1950) courtesy of A&H.

When the 1951 fair opened, construction of the arena and other aspects of Dorton's plan were well under way. The old poultry and industrial buildings near the grandstand and two additional livestock buildings had been removed, and tents with concrete floors housed those exhibits. Exhibitors displayed their cattle in the converted former shops of the State Highway Commission. The youth center, which included a combination cafeteria and recreational center and housed sixty-four girls and sixty-four boys, opened on the western edge of the campus (it was later named in honor of L. R. Harrill, a noted leader of the state's 4-H Clubs). The arena opened partially in 1952, and that same year an "airplane hanger" building for the exhibit of swine was completed.

Finally, the completed arena was dedicated at the centennial fair of 1953. Commissioner of Agriculture L. Y. Ballentine observed that "We are particularly proud to dedicate this year to year 'round service the magnificent new State Fair Arena, rightfully called 'America's Most Modern and Spectacular New Building.'" It was, he wrote, for the people of North Carolina "to use and enjoy every week in the year as a State exhibit center, for livestock shows

and sales, trade shows, conventions, meetings, contests and banquets, and for horse shows, rodeos, circuses and other worthwhile educational, entertainment or inspirational events." With completion of the arena and the youth center, a refurbished grandstand, and upgraded exhibit buildings for the centennial fair, J. S. Dorton's basic concept of the year-round use of the fairgrounds and its facilities could now be fully realized, and all future commissioners of agriculture and state fair managers would oversee the gradual implementation, with significant modifications, of Dorton's grand scheme.

Dorton continued to guide the development of the fair until his death in July 1961, constantly seeking to increase premiums offered to exhibitors; to improve exhibit facilities; to upgrade the vast infrastructure of plumbing, electrical systems, and roads upon which the fair depended; and to do so with funds generated not only by fair revenues but, increasingly, the rental of fair facilities throughout the year. He was unable, however, to make significant strides toward another major building campaign. His futuristic

Like a proud parent, J. S. Dorton poses before the building that now bears his name. The photo dates from after the arena's completion in 1953. Photo courtesy of A&H.

arena, officially named Dorton Arena by the North Carolina Board of Agriculture following his death in 1961, stands as a monument to his vision and has long since become the "unofficial" symbol of the fair.

The building also quickly became a venue for a variety of activities and performances, some of which Dorton hardly could have imagined, that drew audiences from Raleigh

matches featuring heroes and villains and a repertoire of perfectly choreographed acts of bodily mayhem. During the 1960s such featured wrestlers as Haystack Calhoun and Lou Thesz, a legitimate world heavyweight wrestling champion, battled before appreciative crowds, and in the 1970s North Carolina native Rick Flair captivated the crowd with his sweeping, colorful robes, long platinum-blond

Immediately upon its construction, Dorton Arena became the most recognized symbol of the state fair. Fairgoers in front of the arena at the 2000 fair. Photo courtesy of NCDA&CS.

*W*restling became the arena's first successful entertainment series not directly linked to the October state fair, a rather unexpected validation of Dorton's vision of the fairgrounds as a year-round facility.

and the surrounding area. In March 1963, for example, professional wrestling came to Dorton Arena and was an immediate success. By the middle of the decade, the arena regularly hosted wrestling events, and by the early 1970s it devoted every Tuesday evening to professional wrestling bouts that featured some of the biggest names in the sport on the East Coast. Stars of the wrestling circuit entertained enthusiastic crowds with scripted

hair, and flamboyant personality. Wrestling became the arena's first successful entertainment series not directly linked to the October state fair, a rather unexpected validation of Dorton's vision of the fairgrounds as a year-round facility.

In addition to the sport of wrestling, the Raleigh Ice Caps, a professional hockey team in the East Coast Hockey League, played at Dorton in 1991 and 1998. Likewise, thousands of North Carolinians have fond memories of attending other shows, such as the Barnum and Bailey Circus or Disney on Ice, that played at Dorton until the opening of the Raleigh Entertainment and Sports Arena (now the RBC Center) in 1997. Rock fans recall numerous "big name" concerts at Dorton, such as Led Zeppelin in 1975 and Kiss in 1976.

The Jim Graham Era

Following Dorton's death, L. Y. Ballentine, commissioner of agriculture since 1949, served as fair manager from 1961 through 1963. Ballentine essentially followed Dorton's blueprint for the fair, making few substantial changes, the most important of which was the expansion of the fair from five to six days in 1961. With Ballentine's death in July 1964, the department and the fair came under the control of a man who would shape their destinies for almost four decades, James Graham of Rowan County.

Unrelated to the two William Grahams who had served as commissioners of agriculture in the early twentieth century, Jim

Graham was born on a farm near Cleveland in 1921. He attended Cleveland High School, where he served as president of the school's Future Farmers of America chapter, and went on to graduate in 1942 from North Carolina State College, at which he met a number of influential North Carolinians and set his sights upon one day becoming commissioner of agriculture. After graduation he taught vocational agriculture in Iredell County, then with the help of Commissioner W. Kerr Scott became director of the department's Ashe County Test Farm, a position he held for six years. While there, he also managed the Dixie Classic Fair at Winston-Salem and was chosen to head the North Carolina Hereford Association, positions that extended his knowledge of the state's agriculture and politics. In 1957 he became manager of the state Farmers' Market in Raleigh and continued to improve his political connections as he pursued his goal of becoming commissioner. Following Ballentine's death in 1964, Gov. Terry Sanford, whom Graham had supported in his successful 1960 campaign, appointed Jim Graham to the position for which he so longed and for which he had so long worked and trained.

Commissioner of Agriculture Jim Graham, a large man with enormous energy, carried out an ambitious building program during his thirty-six-year tenure. During fair week, he could always be found on the fairgrounds, meeting and greeting fairgoers from across the state. Here he chats with consumers of pork products at the fair, ca. 1978. Photo courtesy of NCDA&CS.

Talented and dedicated fair managers supported Commissioner Graham in his continuous efforts to enlarge and enhance the fair. Together, Arthur (Art) Pitzer and Sam Rand oversaw the construction of most of the major facilities at the fairgrounds. Pitzer served as fair manager from 1965 to 1982.
Photo courtesy of *N&O*.

With the possible exception of J. S. Dorton, no other individual influenced the development of the modern fair as significantly as did Jim Graham. Since the fair was a division within the Department of Agriculture, Graham left the details of staging the fair to a series of exceptionally capable fair managers who headed the division—Arthur Pitzer from 1965 to 1982, Sam Rand from 1983 to 1997, and Wesley Wyatt, who became fair manager in 1997 and continues in the position as of this writing. But Graham, who clearly loved the fair and often visited other state fairs (especially the large fairs of the Midwest and Texas) with his fair manager in search of ideas to improve and enhance it, shaped the overall strategy for the state fair's development during his tenure in office. Yet, in many ways, Graham only carried to fruition the plans for the fair that Dorton had outlined in the late 1930s.

To the average fairgoer, without question Graham's major contribution was the expansion and upgrading of the fair's facilities. Graham relied upon and expanded Dorton's dictum that the fair generate its own operating budget. He employed political skills and contacts developed over two decades of service in the state's agricultural-political bureaucracy, as well as the competence and connections of his fair managers, to persuade the state legislature to appropriate funds for major building projects (which were invariably named for politicians who made their funding possible). Graham's remarkable building campaign is his

Sam Rand, fair manager from 1983 to 1997. The fair's renovated grandstand facility bears his name.
Photo courtesy of NCDA&CS.

legacy, a remarkable testament to his love of the state fair, his extraordinary political acumen, and the exceptional length of his tenure in office.

Graham began with minor improvements to the fairgrounds during the 1960s, improvements that reflected his belief that the fair should continue to educate and that fairgoers should expect modern conveniences. In 1966 the Children's Barnyard, located at the western edge of the campus, opened. Essentially a petting zoo, the facility allowed children from the state's rapidly expanding urban centers to both see and touch common farm animals. Animal exhibits there also included the young of the species—piglets and calves, lambs and baby goats, ponies, and chicks and ducklings. While clearly a sentimental appeal to the natural human affection for things warm, fuzzy, and alive, the extremely popular Children's Barnyard nevertheless helped link urban children to the state's rural past. While it spared young fairgoers the grim realities of modern animal husbandry, it helped acquaint them with the fact that their milk, meat, and eggs were not manufactured at the supermarket. Two years later the fair added modern rest rooms in a facility located adjacent to the Education Building, one of the fairgrounds' original structures—an expansion of facilities always appreciated by fairgoers, adults and children alike.

As Graham consolidated his political base, his ability to obtain legislative funding improved, and the decade of the 1970s saw an astounding expansion of the fair's facilities. One of Graham's greatest achievements was obtaining land to ensure that the fair would have room for future expansion. In 1978 he and fair manager Arthur Pitzer convinced the state legislature to appropriate $1.25 million for the acquisition of an additional 144 acres. The tract the Department of Agriculture purchased lay immediately to the west of the fairgrounds; it brought the total amount of land owned by the state for use by the state fair to 344 acres. That additional acreage presently has yet to be developed, but it gives fair officials the luxury of being able to exercise careful consideration in planning how the property will be used to meet future needs.

During the 1970s and early 1980s, Graham and Pitzer were especially adept at persuading the legislature to fund an expansion of the fair's exhibit space. They succeeded in part because they enjoyed bipartisan legislative support, as well as the backing of both Democratic and Republican governors. Moreover, the exhibit space created was designed not only for use during fair week, but also for year-round use by a variety of clients. In a sense, Graham was gradually putting into operation Dorton's ideal of the fairgrounds equipped with facilities that could be used throughout the year. Of course, such use also brought with it the opportunity to generate year-round revenue.

Twins at the Children's Barnyard (2002). Photo courtesy of NCDA&CS.

The decade began with the opening of the Administration Building in 1970, for the first time giving the growing professional staff of the Department of Agriculture's State Fair Division a permanent home on the fairgrounds. With its completion, the people who ran the fair could plan, organize, and administer it with vastly increased efficiency. More significantly, they could now more effectively administer the year-round use of fair facilities. Among other, smaller units constructed during the decade were a Flower Show Building and a greenhouse, both opened in 1974. Floral exhibits had always been a popular feature of the fair, and the new show building recognized their continuing significance, while the greenhouse provided exhibit space both for commercially grown plants and ornamental plants grown by individual exhibitors. The Old Farm Machinery Building, located along the western edge of the grounds, opened in 1974, as did the Ballentine Building. The former, which serves the function its name implies, added a physical dimension to a post-World War II phenomenon: the sense of nostalgia felt by an increasingly urban North Carolina population for its earlier agricultural heritage.

The three most important buildings constructed during the 1970s, each named for a political figure who had contributed significantly to the growth of the state fair, substantially increased and improved the fair's exhibit space. Funded by the legislature during the governorship of Robert W. Scott, the Gov. W. Kerr Scott Building, named for Robert Scott's father, was completed in late 1973. With 22,128 square feet of space on two floors, the Scott Building gave the fair a modern exhibit hall for commercial exhibits and

additional conveniences for fairgoers, including food service and always-welcome additional rest rooms. Within this expansive facility, commercial exhibitors could set up booths from which they might, and did, hawk encyclopedias and vacuum cleaners, easy chairs and electrical appliances, boats and bullets, and a range of household products. Trade associations, too, opened booths at which they attempted to convince fairgoers of the benefits of a variety of agricultural items, including beef products, traditional pork products such as sausage and liver pudding, milk, eggs, honey, peanuts, and yams. Trade associations served samples that a grazing herd of fairgoers devoured and enticed potential customers with a blizzard of discount coupons and entry blanks for "free" prizes. After fair week, the Gov. Scott Building, which seats up to twelve hundred people for banquets and as many as two thousand people before a portable stage for conference presentations, is available for a seemingly endless parade of trade shows and conventions, providing businesses and trade associations an excellent venue in which to display and promote their products and

The State Fair Administration Building, opened in 1970, houses the office of the fair manager and several members of the manager's staff. Its small size makes it impossible for the entire fair staff to be housed there. Photo (2002) courtesy of the author.

The Gov. W. Kerr Scott Building, opened in 1973, houses many of the fair's commercial exhibits as well as the fine arts competition. Photo (March 2003) courtesy of Blankinship.

The Jim Graham Building, which opened in 1976 with more than 100,000 square feet under one roof, is the fair's premier livestock arena. Photo (March 2003) courtesy of Blankinship.

services. Since its opening, the Gov. Scott Building has housed trade shows featuring Civil War relics, computers, pianos, gems and minerals, and collectibles ranging from antiques to NASCAR trading cards and memorabilia. It has accommodated the Southern Ideal Home Show, high school proms, cat shows, craftsmen fairs, and private parties.

The Jim Graham Building, opened for the 1976 fair, provided a facility that especially pleased the commissioner, an old cattleman: a modern livestock exhibit space built specifically for livestock shows held during fair week. The Graham Building replaced badly outmoded livestock exhibit facilities that had been used since the revival of the fair in 1928. The new facility was 100,000 square feet in size, the vast majority of which is open exhibit space on a concrete slab. It houses a restaurant, two offices, a 240-seat sales arena, rest rooms, and an outside wash area for livestock. Like the Gov. Scott Building, the Graham Building is used year-round by a variety of clients, all of whom, of course, pay for the privilege. Designed with a large, open floor space, the building is used primarily by groups and organizations staging shows that encourage those attending to stroll freely among a variety of exhibitors. Clients who use the Graham Building throughout the year reflect the multifaceted interests of the state's citizenry. Typical attractions staged there include bass and saltwater fishing shows, boat shows, gun shows, lumber auctions, equipment auctions, dog shows, goat shows, recreational vehicle and camping shows, Christmas arts and craft shows, classic car shows, and such events as the Southern Farm Show, the National Elementary Wrestling

Championships, the Dixie Deer Classic, the Southern Ideal Home Show, International Festivals, and Kids' Expos.

Also open for the 1976 fair was the Gov. James E. Holshouser Building, named for the sitting governor, the first Republican to hold that office since Reconstruction. Designated the crafts pavilion, the Gov. Holshouser Building is the major facility in the state fair's "Village of Yesteryear"; during fair week it houses a variety of North Carolina and southern craftsmen demonstrating their skills and presenting their crafts. The Gov. Holshouser Building, a circular structure, is a smaller and more intimate venue than the Graham and Gov. Scott Buildings. With a total capacity of 13,600 square feet, most of which is open exhibit space, the building also can accommodate five hundred people for a banquet and up to one thousand people for a conference. It, too, hosts a variety of events during the months of November through September, many of them crafts-related. Non-fair events include Native American art shows, the Lebanese Festival, the North Carolina Renaissance Faire, gourd festivals, Amish quilts and crafts shows, coin and stamp shows, various dances and social events, and, for animal lovers, cat shows, dog shows, and bird fairs.

The Gov. James B. Hunt Jr. Horse Complex, named for the then-sitting Democratic governor and one of the most astute and successful state politicians of the twentieth century, opened in September 1983, giving the fair a world-class horse facility. Jim Graham had created a political division when he proposed in 1981 that the state legislature fund the complex.

Graham requested $4.3 million to construct a horse-show arena that would seat three thousand spectators and provide stalls for one thousand horses and include an outdoor show ring. Arrayed on one side of the divide were the political/agricultural elite of the state, of which both Hunt and Graham were influential members. Their opponents characterized the complex as a toy for the state's wealthy horsey set, an unworthy and inordinately expensive addition to the work-a-day exhibit space the legislature had previously funded. With Hunt, Graham, and longtime fair manager Arthur Pitzer supporting the complex, its funding was inevitable, and the legislature approved the expenditure. Upon its completion, the complex provided practically all that Graham had requested. Since its

opening, it has housed one of the world's largest all-breed shows during fair week, and it hosts regular horse shows and competitions year-round. In 1989 a restaurant and additional office space were added to the facility, which by that time had, as Graham predicted, become tremendously successful. With the construction of the Gov. Hunt Horse Complex, Graham completed the agenda originally envisioned by J. S. Dorton—the fairgrounds in continual year-round use in displaying the varied industrial, agricultural, educational, and recreational resources of the state and the creativity with which its citizens put them to use.

The latter part of the 1980s and the 1990s saw continued construction, although on a somewhat reduced scale. In 1986 workers

reroofed and painted the grandstand and installed 2,800 molded plastic seats under a covered area. Three years later, the fair added an additional 1,200-seat open-air bleacher area to the facility. A Maintenance Building to house fairgrounds equipment and operations functions opened in 1996 near the Gov. Holshouser Building, and 1999 saw complete renovations of the original Commercial and Education Buildings, which were augmented by an impressive atrium. During fair week, this facility displays exhibits mounted by 4-H Clubs and family and consumer groups. That same year saw the dedication of a new waterfall, located north of Dorton Arena, to replace the old waterfall from 1940 that had stood near the main entrance to the fairgrounds but had been demolished in the early 1960s. Like its predecessor, floodlights illuminated the new waterfall, set in a courtyard with geyser-like fountains. In a touching ceremony, Jim Graham named the new fall for his wife, Helen Ida Kirk Graham, whom he had met at the old waterfall so many years before.

An important part of Graham's building program was the constant expansion and improvement of the infrastructure that supports the fair. Such activity included laying miles of plumbing and providing electricity through a network of both underground and above-ground cables. Perhaps the infrastructure most appreciated by fairgoers, however, was the paving of the fairgrounds' roadways. Prior to the Graham era, with the slightest downpour a thick, red mud became a feature of the fair. Rain soaked the 1928 fair, the initial one held on the new grounds, turning the midway and exhibit areas into a sea of mud. According to the *News and Observer*, fairgoers took the mud in stride: "It is a curious fact," an editorialist observed, "that while men and women going to work might have acquired quite a vicious grouch over muddy street crossings and sloppy sidewalks, they put on a wide grin and walk right into the red ooze that paved the midway at the fair after Tuesday's rain."

Some paving projects had begun in late 1959 and 1960 under J. S. Dorton's management, especially after the fair experienced 4.5 inches of rain in eight days during its 1959 stand. The downpour turned the grounds into a quagmire, causing the cancellation of some exhibits. Prior to the 1960 fair Dorton spent $72,000 paving areas around the major

On numerous occasions, heavy rains have turned the fairgrounds into a sea of mud. Here a woman walks, shoes in hand, at the rain-soaked fair of 1947. Photo courtesy of *N&O*.

exhibit halls, the arena, and the grandstands. He also paved some of the fair's roads and graded and covered with crushed stone some of the access roads to the fair's entrances and parking lots. Mud remained a problem in wet weather, however, and it was during Graham's tenure, with the considerable revenues the fair generated, that fair managers conquered it. The broad thoroughfares from the main gates and past the major exhibit halls were paved, as were the narrower streets that snaked through the midway and exhibit areas behind the grandstand, turning the fair's roadways into broad promenades filled with humanity, even after the most furious downpour.

The Graham years also saw significant changes in the fair's operation. Under the direction of fair manager Arthur Pitzer, the fair in 1969 went to a nine-day format, providing fairgoers three additional days to attend the event and the fair's management with the added revenues those three days produced. For the first time in its history, the fair attracted more than a half-million paying visitors. That milestone was achieved in part because of Pitzer's decision to stop mailing thousands of free passes to county officials, who in turn distributed them to schoolchildren and hundreds of persons active in various local agricultural organizations. Under Pitzer's new admissions policy, everyone coming through the gates paid the required admission fee, except for exhibitors, those sixty or older, and children under the age of twelve—all of whom received reduced rates. Gate revenues alone jumped to more than $300,000 that year, and within a decade the fair was attracting in excess of 650,000 paying visitors annually and collecting more than a half-million dollars in gate receipts. While the method of counting fairgoers varied over the following years, attendance grew steadily. By 1990, the fair was attracting more than 700,000 people, and in 2000, blessed with ten days of beautiful, sunny weather, the fair set an all-time attendance record of 846, 724.

Far more significant, however, was the desegregation of the fair, which itself contributed significantly to the increase in attendance. Through 1963 the fair continued to operate on a segregated basis, and in the 1964 premium list the specially designated Category E of Division One remained set aside specifically for exhibits by Negro Home Demonstration and 4-H Clubs and such organizations as the Negro Farmers of America. The 1964 Civil Rights Act outlawed segregation in places of public accommodation and authorized the federal government to withhold funds from segregated public facilities. In 1964, for the first time since the state of North Carolina had acquired it, the state fair was fully opened to all the state's citizens, whether as simply fairgoers or as vendors and exhibitors. Category E of Division One, like many other segregationist practices, simply ceased to exist.

As previously mentioned, during Jim Graham's tenure as commissioner the fair had become a year-round operation. During every month except October, when the fair's facilities were being prepared for or were actually housing fair exhibits, all the fairgrounds' major exhibit halls housed a dizzying variety of trade shows, conventions, fairs, and associational meetings. Fairgrounds facilities also hosted such events as the Special Olympics, cheerleading camps, meetings of state governmental agencies and agricultural organizations, family reunions, and private parties. Dorton Arena emerged as a major entertainment venue, staging spectacles and spectaculars of every sort, from rock concerts to

wrestling matches, from Christian crusades and gospel sings to circuses and wedding shows. By the end of the twentieth century, the Gov. Hunt Horse Complex every month of the year boasted a schedule of horse shows and sales featuring virtually every major breed of saddle and harness horse and pony—paints, Arabians, Appaloosas, Morgans, walking horses, quarter horses, rodeo horses, draft horses, and hunters and jumpers. In addition, the complex hosted a variety of other events, including the Amran Shrine Circus, the Christmas Cowboy Rodeo, and Renaissance and Medieval fairs.

The North Carolina State Fairgrounds Flea Market, which operates each weekend of every month except October, was initiated by six antique dealers who set out tables at the fairgrounds in 1972. By the end of the twentieth century the market had become an enormous commercial enterprise. Presently occupying the fair's original Commercial and Educational Buildings, as well as two geodesic domes, and spilling out onto the fairgrounds' major thoroughfare, the flea market offers bargain hunters the wares of nearly three hundred permanent dealers, as well as those of an equal number of temporary dealers. Advertising itself as the "biggest and best" flea market in the Southeast, it draws dealers from

Commissioner Graham talks with a young couple that has won stuffed animals at some of the midway's games of skill. Photo (2000) courtesy of NCDA&CS.

Above: The State Fairgrunds Flea Market is by far the most successful example of the year-round use of the fairgrounds. Except during October, every weekend the flea market entices thousands of bargain hunters. Shown here are flea market stalls outside the Scott Building. Photo (2003) courtesy of the author.

as far away as Florida and New Hampshire in search of customers for a mind-boggling array of products, clothing, tools, household appliances and supplies, household pets, candies, sporting goods, decorative items, arts and crafts, and antiques. In many respects the flea market is a variant on the old commercial fairs of the colonial period—a spectacle of hawkers and gawkers, of shoppers curious and committed. An attraction unto itself, it is not quite what Dorton had imagined; but it is a spectacle, and for the fair a profitable one, with which he would have been pleased.

At the close of the twentieth century, Graham presided over a complex that represented the completion of Dorton's vision. The state fair remained North Carolina's largest display of the business of agriculture and the life-style of those who engaged in it. Fair exhibits reflected the work of farmers who tilled the land, tended the state's increasingly large herds of cattle and swine, and managed one of the nation's largest poultry industries. While the October fair no longer served as a major venue at which manufacturers and sellers of

The State Fairgrounds Flea Market, opened in 1972, is now one of the largest on the East Coast, with as many as six hundred dealers offering their wares to the public. In this photo, shoppers stroll through the market, with Dorton Arena in the background. Photo (2003) courtesy of Blankinship.

Whatever the bargain hunter seeks, from antiques to zucchini, can probably be found at the flea market at some time during the year. These flea market stalls are located outside the Commercial and Education Buildings. Photo (2003) courtesy of the author.

agricultural implements displayed their products (especially expensive machines such as tractors, harvesters and combines, cotton pickers, and tobacco harvesters), a variety of less-expensive equipment—everything from bulk barn tobacco curing systems to chain saws—continued to be on exhibit to farmers from every corner of the state. The state fair remained North Carolina's most popular and largest social event, attracting nearly a million fairgoers during its annual ten-day run, many of whom were urban dwellers who came for the thrill of the midway rides, to enjoy the grandstand shows, and to get in touch with their rural past. The end of the twentieth century presented Jim Graham with a dramatic moment to make his exit, and he took it, announcing that he would not stand for reelection as commissioner of agriculture in 2000.

In the political struggle that followed Graham's announcement, Meg Scott Phipps emerged victorious, besting her Democratic rivals in May's primaries and defeating her Republican opposition in the November 2000 general election. In January 2001 she took office, succeeding a man who had held it for thirty-six years and who could have easily been reelected, had he chosen to run. Upon taking office, Phipps continued a long tradition, for she is the granddaughter of W. Kerr Scott, whose campaign during the Great

Depression to make the fair a division of the Department of Agriculture may well have saved it, and the daughter of Robert W. Scott, who as governor from 1969 to 1973 supported Graham's efforts to make the fair one of the nation's best and to allow it to serve the people of the state throughout the year. She inherited a state fair under the immediate direction and supervision of a highly professional and experienced staff led by state fair manager Wesley Wyatt, who has served in that position since 1997. In June 2003, Commissioner Phipps resigned her office, an event unlikely to significantly affect the future of the fair. What shifts of emphasis; what expansions, additions, and changes it will see—the future will determine. That it will remain an integral part of the lives of North Carolina's future citizenry, as it lives in the memories of present-day North Carolinians, is a certainty.

While circumstances will dictate precisely how the state fair develops in the future, Wesley Wyatt and other members of the fair's full-time staff of thirty-five employees have continued to improve the fairgrounds and its facilities. Wyatt and his staff have renovated the original Commercial and Educational Buildings, returning the terra-cotta parapet roof to the Hillsborough Street- and Blue Ridge Road-facing sides of the structures.

Meg Scott Phipps, elected commissioner of agriculture in 2000, is the daughter of Gov. Robert W. Scott and the granddaughter of governor and U.S. senator W. Kerr Scott. The commissioner is pictured here opening the 2002 fair. Standing with her are 4-H Club members celebrating the centennial of 4-H. Photo courtesy of the author.

That work included restoration of the deeply carved, colorful columns outside the front doors, as well as on the opposite side facing the fairgrounds. Wyatt has also enclosed the area between the Commercial and Educational Buildings, creating an attractive, airy, well-lighted atrium.

The fair's professional staff, working with the support of the Department of Agriculture, has also created a master plan that they hope will guide future development. In the fair of the near future, the racetrack—for almost the fair's entire existence a centerpiece of fairgrounds activity—will be gone, except perhaps for the grandstand straightaway portion. The midway, increasingly the fair's primary attraction for thousands of visitors, will expand to occupy the acreage on which the racetrack currently stands. Fair management has also proposed and hopes to see another major exhibit space constructed on the western edge of

the grounds. Such space is badly needed—and not just for the October fair. During fair week, management must turn away many would-be exhibitors; but the space is even more badly needed for trade shows, commercial exhibitions, and other activities held during the remainder of the year, as the fair's current exhibit facilities simply do not provide the square footage required to meet the demand. In the not-too-distant future, it is entirely possible, even highly probable, that year-round activities will dwarf the ten-day-long state fair both in terms of revenues generated and time devoted to their management. But come each October, those activities will cease, all hands will turn to producing the venerable autumn festival and spectacular, and for ten days North Carolinians from all parts of the state will again crowd into the fairgrounds to enjoy one of the state's most cherished institutions.

Wesley Wyatt has served as fair manager since 1997, continuing the tradition of constantly improving the fair's infrastructure. Wyatt, shown here at the 2002 fair, with the help of his professional staff, has drafted a comprehensive plan for the fair's future development.
Photo courtesy of the author.

Step Right Up, Ladies and Gentlemen

The Fair as Carnival and Circus

The Origins of the Carnival

Although the North Carolina State Agricultural Society created the North Carolina State Fair to promote the development of scientific agriculture and industry, the fair was never exclusively a collection of agricultural and industrial exhibits. The Agricultural Society recognized the need for entertainment to attract crowds to its educational exhibits and activities. Its leaders also understood the citizenry's desire for such entertainment, especially among those who lived in rural areas, in which entertainment other than that which they themselves provided was nonexistent. Thus, from its inception, the state fair was meant to be fun for those who attended it.

The society began to sponsor "Special Attractions" to lure crowds in the 1880s. In 1888, for example, a much ballyhooed balloon ascension ended in failure and the departure of the balloonist under cover of night. Undaunted, the society sponsored a successful ascension in 1890, made the more spectacular by the balloonist's return to earth by parachute. Thereafter, the fair was incomplete without a balloon ascension

Night rides on the Strates Shows midway at the 1996 fair. Photo courtesy of NCDA&CS.

Secretary Nichols After the Hootchee-Koochee Girls.

claimed to perform only the authentic dances of their respective native lands. Some fairgoers, however, believed the "hootchee-koochee" show to be beyond the limits of good taste and decency and complained to the society, which closed the show. Visitors to the fair with less-developed sensibilities, on the other hand, enjoyed the dancers enough to demand repeat performances. The society entered into negotiations with the show's management, resulting in a "compromise" that saw the show reopen the following day. The society's secretary, John Nichols, a noted Raleigh printer and Republican politician, policed the performance to ensure that the girls were adequately clad. Although the society announced no definition of adequacy, Nichols's observations evidently convinced him that the young ladies remained presentable, and the show continued to attract an audience filled with inquiring minds.

The shows of Victor D. Leavitt dominated the midway during the late 1890s. Among the attractions offered by the Leavitt shows at the fairs of 1897 and 1898 were the "Crystal Maze"; Buckley's Great Horse Show; Mont Morenzo's

Trout of the Moors; Jim Key, billed as America's smartest horse; and several other lesser attractions of the time. At a time when the Hearst and Pulitzer newspaper chains were bombarding the American public with headlines proclaiming the brutal treatment the liberty-loving people of Cuba were receiving at the hands of their Spanish masters, the "Starved Cubans" also appeared. (While such displays provided a popular commentary on American foreign relations, they also aided Hearst, Pulitzer, and their supporters in successfully spurring the nation into a war with Spain.) Numerous smaller shows and independent attractions likewise played to fair audiences of the late 1890s. By the turn of the twentieth century, the modern midway, with its freak shows, girlie shows, thrill shows, featured attractions, and games of chance had become an integral part of the fair. While they undoubtedly drew larger crowds to the fairs, they also diminished the fair's value as an instructional institution.

The festive air of fair week, the large crowds, and the money fairgoers brought with them combined to create problems for the State Agricultural Society. The fair presented the ideal opportunity for the practitioners of certain criminal arts. Pickpockets appeared at the antebellum fairs and remained troublesome throughout the nineteenth century. Counterfeiters began to work the fair as early as 1873. Drinking and gambling, however, posed the most serious threat to the harmonious operation of the fair. The society sought to regulate both evils through the efforts of Raleigh policemen and specially deputized fair marshals and on the whole did so with considerable effectiveness.

Soon after the fair's inauguration, the Agricultural Society was forced to develop a policy on gambling, and in 1856 it responded by forbidding all games of chance. In the early post-Civil War years, however, gambling concessions of various descriptions operated at the fair despite the society's best efforts to prohibit them. In 1874 the society relaxed its position and instead sought to regulate gambling, licensing roulette and other wheel-game operators, a decision roundly condemned by the local press and pulpit. In 1875, capitulating to its critics, the society returned to the policy of prohibiting gambling, which continued until 1885. In that year the society again adopted a policy of licensing "games of chance" but continued to forbid "skin games" —those which, presumably, the fairgoer had little "chance" at all of winning. In 1891 the state legislature sought to cope with the problem, passing a bill forbidding "all games of chance, wheels of fortune, and gambling of all species at any fair. . . ." The 1891 bill created a

recognized the fairgoers as easy marks and continued to fleece them. Fair officials shut down several games at the 1892 fair, and in 1899 a number of game operators were arrested on the midway.

The Carnival of the Early Twentieth Century

Carnivals and shows offered at twentieth-century fairs, like those of the previous century, reflected the changing mores, values, and preoccupations of the larger society. The organized entertainment provided by professionals bidding for the interest, and the money, of fairgoers gradually evolved into two distinct forms—the carnival and the grandstand show. Both underwent significant modifications in the twentieth century. Consistently, however, carnival operators and show-business impresarios attempted to provide fairgoers the thrills and chills, the moments of elation and anticipation, which kept them spending their hard-earned dollars.

The 1900 fair featured a variety of amusements. "Professor" J. T. Cook urged fairgoers to win prizes at his games of chance, rather than spend their money to see "Basco, the original snake eater." Inasmuch as the American soldiers were then engaged in guerrilla warfare against Filipino insurgents, fairgoers could pay to see the "enemy"—Hai-low, the Filipino boy "who looks like a monkey, acts like a monkey, lives on raw heads and bloody bones," and was descended from a "a race of

By the turn of the twentieth century, the modern midway, with its freak shows, girlie shows, thrill shows, featured attractions, and games of chance had become an integral part of the fair. While they undoubtedly drew larger crowds to the fairs, they also diminished the fair's value as an instructional institution.

state of open hostility between fair officials and game operators. Fair officials who licensed gamblers illegally, or who failed to stop any gambling they observed, could be forced to reimburse the fairgoer for any money lost while gambling and to pay an equal amount to the state's School Fund. Directed more at fair officials than at operators of games of chance, the law failed to deter gamblers, who

blood drinkers and flesheaters." On the midway, in addition to attending freak shows or a "snake menagerie," the fair visitor could ride the Ferris wheel, the "Crystal Maze," the "Tom Thumb Rail Road," the "Aerial Ocean Wave," or "the latest Coney Island novelty, the Razzle Dazzle." The fair also featured a "regular" circus with acrobats, trapeze artists, and twenty trained dogs, in addition to a balloon ascension and a parachute jump. The fair closed with a fireworks display.

A decade later, little had changed. Fair attractions included Buckskin Bob's Wild West Show, the Guthries, high-stilt walkers, trapeze artists, a high-wire walker, Cameroni, "The Great Teeth Descensionist" who held to a slide on a wire with his teeth, and Frank Coleman, "World Renowned Balloon Artist." In the Rollins Wild Animal Show, fairgoers could watch as "men go in cages with wild beasts and make them kneel at their feet." A freak show with an educated horse, an eighteen-year-old fat girl who weighed eight hundred pounds, and other "oddities" likewise played the fair, along with an early version of Harry Kaplan's Dancers, a show made up of girls who, the *News and Observer* assured its readers, were "of the highest class and dance with grace and ease."

Daredevils and Speedsters

Once Wilbur and Orville Wright flew their biplane off the sands of Kitty Hawk in 1903, it was inevitable that someone would turn the flying machine into a carnival act. By 1906, intrepid airmen were awing crowds at state fairs in the Midwest, and well before the First World War they had become standard attractions. At the 1915 North Carolina State Fair, for example, as the war raged in Europe, a Captain Worden bombarded a "fort" from his aeroplane twice daily, to the delight of the crowd. In 1920 four airplanes piloted by "army aviators" appeared at the fair, reflecting the rapid advances in aviation resulting from the war. But airplanes were difficult machines to maneuver close to large crowds. Air shows required huge amounts of space; the automobile, however, was made for the grandstand track. Thus, the stunt driver

soon superseded the stunt flyer. At the 1920 fair an Overland automobile staged a fourteen-foot ramp-to-ramp jump, a feat with which all automobile owners could identify. The stunt driver proved to be a precursor of the organized automobile thrill show, which would dominate grandstand events from the 1930s through the 1970s and continue to thrill fair crowds until the late 1990s.

Well before the automobile thrill show became a regular feature of the North Carolina State Fair, however, another type of automobile daredevil—the race car driver—had become a standard attraction. As soon as automobiles appeared, men began to race them. Originally, motorcar clubs in larger cities sponsored races, which were basically entertainment for the very rich. In 1902, however, nine auto clubs met in Chicago to form the American Automobile Association (AAA). In 1909 the AAA formed its Contest Board and began sanctioning automobile races, especially on the East Coast. State fairs, with their horse tracks, provided a natural venue for automobile races, and they were soon staged under Contest Board sanction. Former drivers who had discovered a less-hazardous means of making a livelihood usually promoted such races. By the early 1920s open-wheeled Indianapolis-type cars were racing at the North Carolina State Fair under the promotion of Ralph A. Hankinson, a noted driver of the pre-First World War era, who billed his firm as the "foremost 'pro-motors' of dirt-track races in America."

In 1928 Hankinson and the AAA-sanctioned races returned to the North Carolina State Fair for two days. According to the *News and Observer*, "Thousands lined the track around the upper end and filled the grandstands, watching a score or more of the foremost racers in the country compete for purses." Drivers included Bob Robinson, "the acknowledged ace of American half-mile tracks," and Doug Wallace, the 1927 southern dirt track champion. Robinson, driving a Frontense, thrilled the crowd with a record eleven-minute, sixteen-second finish in a ten-mile race, while Herman Schurch, "speed merchant from Hollywood, Cal.," recorded the fastest lap at 30.4 seconds.

The fair continued to feature AAA-sanctioned races until well after the Second World War. During the early 1930s, Eddie

One of the earliest motorized stunt shows to play the state fair was the so-called "Globe of Death." The show featured motorcycle riders performing inside a steel globe, such as this one, which appeared at the 1934 fair. Reproduced from the 1934 *Premium List.*

THE

NORTH CAROLINA STATE FAIR

PRESENTS

LOOPING NIXES

Thrilling Motor Cycle Loop the Loop, in Globe of Death. Afternoon and Night Grandstand Performance

93

From the initial contests of the early twentieth century, automobile racing at the state fair featured open-wheeled, Indianapolis Speedway-type cars. Hankinson Circuit drivers competed at the fairs of the 1920s and 1930s. Shown here is an advertisement for auto races at the 1934 fair. Reproduced from the 1934 *Premium List*.

Rickenbacker, one of the most colorful characters in American automotive and aviation history, promoted the races. By the First World War, he was one of America's most famous drivers and held several speed records. During that war he became America's top "air ace" with twenty-two confirmed "kills" of enemy airplanes. After the war, with the sanctioning power of the AAA Contest Board behind him, Rickenbacker entered the business of promoting automobile racing.

After the Second World War, Sam Nunis, general manager of Nunis Speedways, emerged as the major promoter of automobile races at the North Carolina State Fair. A competitor of the legendary Bill France, founder of NASCAR, Nunis operated several East Coast racing venues. By 1953, still operating under AAA sanction, he was supplying what were essentially open-wheeled sprint cars to the state fair. Nunis continued to promote automobile races at the fair until 1967, the last year "Indy"-type car races were held there. In 1969 the fair briefly replaced the Indy cars with a NASCAR-approved grand touring auto race. Such racing at the fair could not, however, compete with the appeal of rapidly expanding NASCAR stock-car tracks, including several in North Carolina, and auto racing soon disappeared from the state fair.

Below: Nunis continued to promote open-wheeled racing, although the cars were more streamlined and faster. Photo courtesy of NCDA&CS.

Sam Nunis, who rivaled NASCAR founder Bill France as a promoter of auto racing on the East Coast in the 1940s, promoted such racing at the fairs of the 1940s and 1950s. Photo courtesy of NCDA&CS.

George Hamid: A Showman Changes the Fair

By the time the North Carolina State Agricultural Society's financial collapse placed the fair under state control in 1928, large, well-organized organizations contracted for the right to supply the midway with its carnival. In 1928 and 1929 the Greater Sheesley Shows, "America's Greatest and Cleanest Traveling Amusement Promenade," played the fair, giving visitors "a carnival with real shows, rides, concessions and unusual exhibits." Rides included the "Coney Island Flyer," the "Whizzing Zip," and the "Giant Caterpillar," along with such old standbys as the Ferris wheel and the carousel. Male fairgoers were probably more enticed by the Sugar Cane Revue, an early "girl show," or the Autodome, a motorcycle thrill show, while the family might have attended the Monkey Circus or the Animal Freak Show or have gone to see the Vampire. Arriving in Raleigh in forty "big railway cars," the Harry Melville-Nat Reiss Shows played the 1930 fair, with "rides whirling, clowns clowning, freaks freaking, and side shows showing." Among its seventeen attractions were Whoopee, "with all its name implies"; Get Happy, "an elaborate, tuneful, mirthful Colored Revue"; Superba, "a mystifying Spectacle"; the Fat Family, "Tons of talent and laughter"; and The Unborn, "human life in the making." The carnival also featured a midget family, circus sideshow freaks, a rodeo, and a monkey circus. The Melville-Reiss Shows boasted ten rides, including the "Skooter," the "Lindy Loop," the "Whip," and the "Merry Mixup."

During the late 1920s a showman who would leave an indelible imprint on the North Carolina State Fair arrived on the scene in the

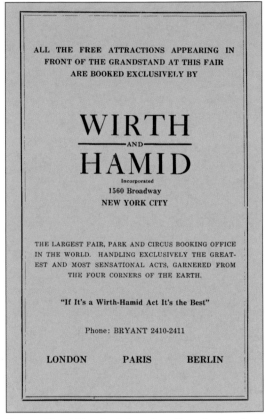

George Hamid booked the attractions featured at the 1928 state fair. This advertisement from the fair's premium list from that year suggests Hamid's flair for showmanship. Reproduced from the 1928 *Premium List.*

person of George Hamid. Hamid had literally grown up in the rough-and-tumble world of an industry noted for its cutthroat competition. Born in Lebanon in 1896, Hamid set sail for America in 1906 with two cousins, hoping to join their Uncle Ameen in America. In route to the United States, the three boys, who formed the basis of an acrobatic tumbling act, stopped in Marseilles, France. There they auditioned for a spot in Buffalo Bill's touring Wild West Show, which they obtained through the intervention of the famed sharpshooter Annie Oakley. Hamid and his cousins toured Europe for the remainder of the season with the troupe, during which time Hamid learned English under the tutelage

The premium list for the 1929 state fair advertised an extensive list of free attractions. Reproduced from the 1929 *Premium List.*

The
North Carolina State Fair
Presents the Following
FREE ACT PROGRAM
October 14-19, 1929

THE GREAT WILNO
The Human Cannon Shell

THE PHUNNY PHORD
The Only Act of Its Kind in Existence

THE BONHAIR TROUPE
The Acrobatic Aces of the Ages

DEMARLO AND MARLETTE
A Circus Sensation of Marvelous Agility

JEAN JACKSON TROUPE
A Thrilling Cycling Act

RITCHIE WATER SHOW
A Trick, Fancy and High-Diving Water Circus

CERVONE'S CELEBRATED BAND
and Vocal Soloist

HORSE RACES
$7,200.00 in Purses

AUTOMOBILE RACES
With Celebrated and Nationally Known Drivers

FIREWORKS
American Fireworks Company of Boston Presents Elaborate
Displays With Change of Program Each Night

Wirth and Hamid, Inc., New York City
Presents **Sensations of 1930**
A SCHOOLEY PRODUCTION

A BRILLIANT AND SPECTACULAR PRODUCTION COMPOSED OF GLORIOUSLY BEAUTIFUL GIRLS

of Annie Oakley. When Buffalo Bill returned to America, Hamid returned with him and joined his Uncle Ameen, with whom he labored as a tumbler at various East Coast venues. At age thirteen, Hamid entered a contest at Madison Square Garden, where he won the "Acrobatic World's Championship." The following year he rejoined Buffalo Bill's show as the head rider with a troupe known as the "Far East Riders" and began an American tour, amazing audiences with his acrobatic tricks on horseback.

Buffalo Bill's show folded in Denver in 1913, and Hamid and his cousins, performing as tumblers, worked their way back to New York. From New York, Hamid's acrobatic troupe began to play Atlantic City, as well as to accept various tour offers in New England and the Midwest. In 1917 Hamid organized a "Streets of Cairo" act for New York's Coney Island, then developed a traveling carnival show based on the talents of a stripper billed as "Fatima." Beginning in 1918 he operated a combined Wild West and Far East show. In 1920 he teamed with Frank Wirth and Herman Bluemfield, an importer of thrill acts, to form the Wirth and Bluemfield Booking Agency. After buying out Bluemfield, Hamid and Wirth expanded their operations and in 1925 became the exclusive talent-booking agent for the Ringling Brothers Circus. In the world of grandstand acts, Hamid was the king of bookings by the mid-1920s.

When the fair opened in 1928 under state auspices following a two-year hiatus caused by the financial collapse of the North Carolina State Agricultural Society, Wirth and Hamid

A GORGEOUS
Broadway Revue
Composed of the sprightliest, daintiest, and prettiest aggregation of dancers that has ever been assembled in one show for outdoor presentation. . . . This Revue carries its own stage, lighting effects, electricians, and special scenic effects erected on the race track.

Ziegfeldian In Its Gorgeousness
THE NORTH CAROLINA STATE FAIR
October 12-17, 1931
RALEIGH, N. C.

supplied the grandstand shows. Hamid's relationship with the state fair continued well into the 1960s, with Hamid applying a formula for grandstand entertainment straight out of the late nineteenth century. Hamid's shows invariably represented a mixture of vaudeville and the circus, a medley of acts chosen to bedazzle and bemuse, but never bore, the audience. The usual mix included acrobatic troupes, jugglers, animal acts, a comedy routine or two, at least one "thrill" act (such as high-wire performers or a human

LARRY BOYD & PHIL WIRTH, Inc.,
NEW YORK CITY

Presents

THE GREAT WALLENDAS

Now in Europe playing return engagement at Olympia Circus, London.
Late feature with Ringling Bros.-Barnus & Bailey Circus.
The last word in high wire features.

cannonball), musical numbers, and, always, a bevy of spectacularly costumed beauties. The Wirth and Hamid show of 1928 was typical of what fair audiences could expect at a George Hamid grandstand show. It included singing acts May Wirth and the Wirth family, the Billy Rice Trio, Cervone's Band, Eddie Lerch's dog

show, and the "Royal Dancing Review." In 1930 Wirth and Hamid presented "Young China, . . . a super-sensational display of Chinese Posturing, Balancing, and Plate-spinning," along with Huling's Sea Lions, straight from the Ringling Brothers and Barnum and Bailey Circus; Cervone and his band; Abe Goldstein, "America's foremost character clown"; and the Comedy La Dells, "a pretty girl and several boisterous clowns, using trick doors and walls to entertain."

Fearing that the worsening Great Depression might bankrupt the North Carolina State Fair and place its members in a position of personal financial liability, the Board of Agriculture in 1933 leased the entire operation of the fair to Hamid. From 1933 through 1936, Hamid was the fair, and as such he supplied all the entertainment, both on the midway and at the grandstands. For the major venues to which he supplied carnivals, Hamid booked Max Linderman's "World of Mirth" shows, one of the nation's largest carnivals. Like the Ringling Brothers and Barnum and Bailey Circus, Linderman's shows traveled by rail. Accompanied by Hamid, who was determined to see that his fair was a profitable one, the Linderman carnival pulled into Raleigh to

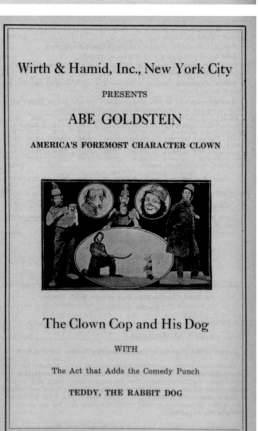

May Wirth and the Wirth family regaled fairgoers at the 1932 fair with feats of bareback riding, always a favored Hamid act. Reproduced from the 1932 *Premium List.*

Grandstand shows always included at least one comedy act. Abe Goldstein performed at the 1930 fair. Reproduced from the 1930 *Premium List.*

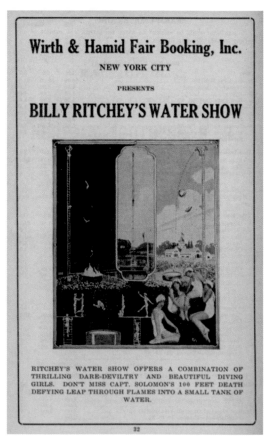

Not all of Hamid's revues were typical chorus line productions. Water-show beauties performed at the 1933 fair. Reproduced from the 1933 *Premium List.*

open the 1933 fair with 19 novelty shows, 3 fun houses, 10 modern rides, and 2 brass bands.

Hamid and Linderman, both experienced showmen, fully understood the ups and downs of the carnival world, but setting up the carnival in Raleigh proved an exceptionally daunting task. They arrived at the fairgrounds to find "a dozen gypsy tents dotting the midway," their occupants encouraged by "a somewhat un-cooperative sheriff and his crooked buddy, the coroner." The sheriff and the coroner had informed the gypsies that Hamid had leased the fair and planned to throw them out. Hamid had the realization that "Obviously, I couldn't call on the law to evict the gypsies, so we tossed them bodily off the lot. . . . We led Max's elephant, Ginger, out to a row of concessions, where I gave the thugs 15 minutes to clear out. Meeting only jeering defiance, I pronged Ginger and she demolished the first tent. The owner rushed at me and I had to flatten him with my fist. The rest of them packed their belongings and fled."

Although he had succeeded in occupying the fairgrounds, Hamid was still not out of trouble. He received death threats, and the sheriff and coroner informed him that they could close down the fair under an 1826 law unless he proved more cooperative. Hamid appealed to Josephus Daniels, publisher of the *News and Observer*, for help. Daniels obtained Hamid an appointment with a superior court judge, and Hamid explained to the judge that the sheriff and coroner had "arrested" their gypsy friends, carried them to the fair's "jail," and extracted their payments before releasing the "suspects." The judge, informed about the operation by Daniels, had the sheriff and coroner in the anteroom, invited them to join his conversation with Hamid, and warned them that if Hamid experienced any more difficulty, they would wind up behind bars.

In his initial year as operator of the North Carolina State Fair, Hamid relied upon his tried-and-true formula in booking grandstand acts. His grandstand show in 1933 featured the usual acrobatic tumbler troupe, an entire circus, an elephant show, a trapeze act, and lots of costumed showgirls appearing in a "New York Winter Garden Review." For the next three years under Hamid, Max Linderman's World of Mirth Shows provided fairgoers a variety of freak shows, girl shows, fun houses, rides, games of chance, and other

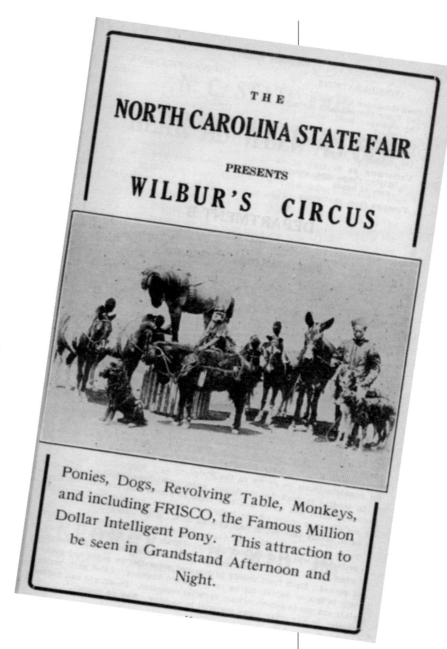

attractions in which to immerse themselves in hopes of forgetting, at least for the moment, that America continued in the grips of its most severe economic depression. Although Hamid continued to procure some original acts for the state fair audience, his grandstand shows relied heavily upon "Circus and Hippodrome" performers.

On January 30, 1937, W. Kerr Scott, the newly elected commissioner of agriculture, notified Hamid that the Board of Agriculture had unanimously voted to cancel his lease on the fair's operation. The notification was a serious loss to Hamid, for the lease had guaranteed him 15 percent of the fair's gross receipts (or at least $8,000), in addition to 50 percent of all net profits above $15,000, as stated earlier. Hamid accepted the loss of the lease but fought to keep his shows at the fair, now a division of the Department of Agriculture

Another of Hamid's major strengths as a booking agent was his ability to obtain top-notch circus animal acts, such as Wilbur's Circus, which played the 1934 fair. Reproduced from the 1934 *Premium List*.

under the management of J. Sibley Dorton. In June 1937 Hamid complained bitterly to Scott that Dorton would not renew the Hamid shows for the 1937 fair, to which Scott replied that the final decision concerning entertainment at the fair was Dorton's. Hamid had reason to fear losing the state fair booking, for as early as November 30, 1936, following Scott's victory at the polls but before he took office, Elbert E. Foster of Mecklenburg County, a member of the North Carolina legislature, had written Scott on behalf of Capt. John M. Sheesley, operator of "Sheesley's Mighty Show," who clearly sought to regain the privilege of booking his carnival for the fair's midway. Foster declared that if Scott would help Sheesley, "we would do our utmost to secure any legislation that you may desire passed by the 1937 legislature in connection with the staging of the State Fair." Meanwhile, Dorton himself was receiving letters in support of the

Goodman Wonder Shows, which, writers assured Dorton, wintered in North Carolina where it spent considerable sums rebuilding its attractions.

For whatever reason, Hamid won the battle of the promoters, and the 1937 fair retained both Hamid's grandstand acts, which included "The Review of Tomorrow," and Linderman's World of Mirth Shows, for the midway carnival. Hamid's concerns that Dorton might book other shows were not unfounded, however. Although Hamid continued to supply the grandstand acts, in 1938 Dorton dropped Max Linderman's World of Mirth Shows on the midway, replacing them with the Johnny J. Jones Shows. Either the Jones shows proved disappointing or Hamid convinced Dorton to resume booking Linderman's World of Mirth Show, for in 1939 the World of Mirth returned to the state fair midway. The Linderman shows played the fair

"Laugh Land" was one of the midway features of the Johnny J. Jones Shows, which played the 1938 fair. Photo courtesy of *N&O*.

through 1941, after which the outbreak of the Second World War suspended the fall classic. Max Linderman died in 1944; but when the fair resumed in 1946, the Linderman shows returned for two more years on its midway.

Stunt Men and Hell Drivers

George Hamid, meanwhile, had discovered a sensational grandstand act and introduced it to the state fair in 1936; it became a permanent feature of the fair into the last decade of the twentieth century. The act belonged to a former automobile test driver named Earl

"Lucky" Teter. Teter was not the first auto daredevil. Barnstorming auto daredevils had toured fairs in the first two decades of the twentieth century, but the organized auto thrill show began in 1923 when B. Ward Beam organized the International Congress of Daredevils in Toledo, Ohio. By the late 1920s, Beam's troupe was playing the big state fairs of the Midwest. Teter's genius was to convince Plymouth Motors in 1933 to sponsor his act and provide him with a fleet of brand new automobiles. In addition to marketing Plymouths, Teter also used his cars to market automobile-related products, including oil, gas, tires, and auto parts. A born showman,

During the 1930s, Lucky Teter transformed the barnstorming auto thrill show into a tightly choreographed display of precision driving stunts. Here Teter performs a barrel roll at the 1938 fair. Photo courtesy of *N&O*.

Teter coming off a ramp at the 1938 state fair. Photo courtesy of *N&O*.

Teter and his Hell Drivers used timing, preparation, and sheer courage to perform precision driving stunts that thrilled crowds at state fairs and other outdoor venues nationwide. In these two photos the charismatic Teter is shown signing autographs at the 1940 North Carolina State Fair. Photos courtesy of *N&O*.

a meticulous planner, and absolutely fearless, Teter developed a reputation for staging spectacular stunts. He recruited a cadre of stuntmen, billed them the Hell Drivers, and wowed audiences at outdoor venues throughout the East and Midwest and in Canada. By the time Hamid brought Teter and his Hell Drivers to the 1936 state fair, Teter was already a huge star.

Teter, "the Babe Ruth of stunt driving," was a man of extraordinary skill and courage, the star of his show as well as its promoter. Later auto thrill-show operators acknowledged him as the man who invented or perfected practically all the basic stunts in their repertoire. Teter pioneered the ramp-to-ramp jump, in which his car hurdled over a variety of vehicles, sometimes stationary, sometimes moving; the T-bone crash, in which a stunt driver deliberately drove his car off the end of a ramp and into another vehicle; the "Roll Over," in which the driver deliberately rolled his automobile as many times as possible; and the High Ski, in which a driver took a car up a single-track ramp, flipped it up on two wheels, and held that position for yards before bringing the automobile down on all four wheels. In their precision-driving stunts, his drivers crisscrossed four cars at high speeds, sometimes bringing two of the cars off carefully placed single-track ramps in the process. Teter's act also featured "clowns," men who

Teter was famed for his ramp-to-ramp jumps of other vehicles. At the 1941 fair, his last, Teter leaps a bus. Photo courtesy of N&O.

stood motionless as cars crisscrossed about them, missing them by inches, or lay between the ramps as cars on either side used them to go into the High Ski maneuver. Teter's grand finale, which he performed himself, was an end-over-end, side-over-side somersault. Crowds everywhere loved him and could identify with the "stock" cars he drove, which were modified only by additional bracing and the relocation of gas tanks to the trunk. North Carolina State Fair crowds were no exception, and Teter returned in 1937 by popular demand. He played every fair through 1941, the last one held until after the Second World War. Teter's luck finally ran out on July 5, 1942, at a benefit performance for the Army Emergency Relief Fund at the Indiana State Fairgrounds. As he approached the takeoff ramp for a ramp-to-ramp jump of a truck, the carburetor on Teter's car malfunctioned. His automobile lost speed, cleared the truck, and then slammed into the underpinning timbers of the landing ramp. He died on the scene.

When the 1946 North Carolina State Fair opened, Jack Kochman's Hell Drivers were there to perform for fairgoers. Kochman's drivers, like Teter's, drove factory-sponsored stock cars—Dodges. Unlike Teter, however, Kochman was a promoter, not a driver. The popularity of the auto thrill show continued to increase after the war as Americans rushed to buy automobiles with money earned in the postwar prosperity the country enjoyed. Joie Chitwood's troupe soon began to play the fairs with Kochman. Chitwood, a native of Oklahoma, was a noted race car driver of the 1930s who joined Teter's Hell Drivers. Following Teter's death, Chitwood purchased his fleet and went on the fair circuit with his own act. By the 1950s, Chitwood had switched to Fords, and his automobiles, along with Kochman's Dodges, became staples of grandstand entertainment at the fair, with both acts usually performing at least twice during fair week, always on alternating days. While both Kochman and Chitwood added

After World War II, renowned showman Jack Kochman, not himself a stunt driver, reassembled the Hell Drivers. Kochman's Hell Drivers performed as the premier grandstand act from 1946 until the early 1970s. Kochman is pictured at the 1952 fair. Photo courtesy of NCDA&CS.

flourishes to their acts, essentially they performed the same stunts that Teter had perfected in the 1930s. In a nation increasingly enraptured with ever more powerful automobiles, both acts drew large and appreciative crowds annually. Kochman's Hell Drivers played the fair until the early 1980s, then made a final appearance in 1988, after which Jake Plumstead's Auto Thrill Show replaced them until 1991. In 1991 Joie Chitwood Jr. brought his fleet of Chevrolet Camaros to the fair and returned every year until 1997. It is fitting that

Chitwood, whose father played the fairs of the 1950s and 1960s with his own auto thrill show and who had driven with Teter at the fairs of the 1930s and 1940s, would bring down the curtain on stunt-driver shows before grandstand crowds. While failing to measure up to their popularity in the immediate post-World War II years, the auto thrill show could still attract a crowd. Fair management, however, deemed such shows too expensive. The so-called "demolition derby," a modern embodiment of the automobile spectacular,

Joie Chitwood drove for Lucky Teter and after World War II played the state fair with his own troupe of stunt drivers. His son, Joie Jr., kept the show on the road and played the fairs of the 1980s and 1990s. Here Chitwood cars are preparing for a crisscross maneuver at the 1983 fair. Photo courtesy of NCDA&CS.

A Chitwood driver coming off a ramp at the 1996 North Carolina State Fair, the last year stunt drivers performed there. Photo courtesy of NCDA&CS.

The demolition derby, begun in 1984, eventually replaced the precision stunt driving shows as a grandstand attraction. They were far less expensive to stage and delivered real violence. Here cars line up for the start of the derby at the 1999 fair. Photo courtesy of NCDA&CS.

A demolition derby driver in action at the 1999 fair. Photo courtesy of NCDA&CS.

introduced in 1984, required little skill on the part of drivers and delivered a tremendous bang for the buck. The demolition derby survived into the twenty-first century as the fair's only form of automotive entertainment.

The James E. Strates Shows and the Modern Carnival

While George Hamid continued to supply the fair's grandstand acts in the post-World War II era, the midway carnival went to another vendor in 1948. That vendor was James E. Strates. Generations of fairgoers regarded his shows an integral part of the fair experience, for the Strates shows played the fair each year without interruption from 1948 until 2001. Indeed, only horse racing endured at the fair as an ongoing organized activity for more years than did the Strates shows, and the shows outlasted horse racing. Indeed, to some extent the enormous popularity of the Strates midway contributed to the demise of horse racing. For those who attended the fair primarily because of its carnival, the James E. Strates Show was the fair for more than a half a century.

Remarkably, James E. Strates's life story parallels that of George Hamid. Strates, a

In 1922 James E. Strates founded the Strates Shows, which soon became one of America's largest touring carnivals. Strates Shows played the North Carolina State Fair continually from 1948 to 2001. Although Strates, pictured here in 1953, died in 1959, his shows continued under the direction of his son, E. James Strates. Photo courtesy of NCDA&CS.

native of Greece, immigrated to the United States in 1909 at the age of fifteen, eventually settling in the Endicott, New York, area. There he made a living working at various jobs, including a stint in a shoe factory. A superb athlete, he learned to wrestle at the local YMCA and in 1919 set out to seek his fortune as a wrestler (under the show name "Young Strangler Lewis") with a carnival that played the southern region of New York. "Strangler" continued as a carnival wrestler in the early 1920s and in 1922 lost a legitimate middleweight world title match. In 1922 he joined forces with two other carnival wrestlers to form the Southern Tier Shows, which performed primarily at venues in lower New York. In addition to the athletic show, Strates's carnival began with a merry-go-round, a Ferris wheel, fifteen concessions, and three sideshows. In 1927 Strates bought out his partners, and by 1927 he was operating the largest carnival in New York State. His carnival, which traveled by truck, had added a variety of freak shows, animal acts, a contortionist, and a knife-throwing act. Hit hard by the depression, Strates kept his carnival on the road and in 1932 changed its name to the James E. Strates Shows. The following year he used his savings to purchase flatbed rail cars and converted from truck to rail transportation. By the end of the decade he was employing twenty-five rail cars and sixty-one trucks and wagons to move his shows. While Strates was wintering in Mullins, South Carolina, in 1945, fire destroyed his entire carnival except for the rail cars used to move it. Strates received a quarter of a million-dollar insurance payment to rebuild his organization and by the end of 1946 had twenty sideshows, twenty-one rides, and a variety of concessions back on the rails. Thus, when the Strates train arrived in Raleigh for the 1948 North Carolina State Fair, James E. Strates was already recognized as one of the nation's leading carnival operators, and his show was one of the nation's largest carnivals, employing nearly three hundred people.

The Strates shows employed a typical carnival layout. In 1948 Strates added a popular burlesque act—or "girl show," as they were known in the trade—called "A Night on Broadway," which starred one of the best-known exotic dancers of the decade, who billed herself as "Georgia Southern." As in other large carnivals that played American fair

circuits, the newly obtained girl show heavily influenced the midway layout. The girl shows always played at the curve of a midway arranged as an open horseshoe, a position selected to draw maximum crowds—especially males, who normally carried more cash and were more prone to spend it—down the sides of the horseshoe. On the sides of the horseshoe immediately up from the girl shows were Strates's other shows—especially on the right side of the horseshoe, for the crowds tended to move in a counterclockwise manner. There fairgoers usually encountered several freak shows and at least one animal or snake show, several fun houses, and other attractions. At the entrance to the sides of the horseshoe stood a variety of games of chance and food concessions. In the middle portion of the horseshoe, creating an open ellipse along which fairgoers paraded, gawking at the Strates attractions, was arranged a series of rides. The entire area, night and day, bombarded fairgoers with a cacophony of sound—music blaring from speakers, the amplified noise of side-show barkers, recorded screams from the fun houses, the constant whir of the enormous machines that propelled their riders through space in a variety of positions. By

night the midway blazed with brilliant neon colors, always in motion, blinking, flashing, rotating, and providing a spectacular, mesmerizing display designed to attract fairgoers' attention long enough to extract their dollars.

By the time Strates died in 1959, his shows had become a central and beloved part of the fair. Strates died on Sunday, October 11, as his

Above: This 1966 aerial view of the fair depicts the typical layout of the Strates Shows (center foreground). Fairgoers streamed into the midway from entrances on either side of the horseshoe-shaped layout, which appears virtually at the center of the photo. Photo courtesy of *N&O.*

Left: Members of the Strates family enjoyed a long-term business and personal relationship with the management and personnel of the North Carolina State Fair. Mrs. James E. Strates, widow of the founder of the shows, is here pictured having dinner at a Raleigh restaurant while in town for the 1960 fair. Photo courtesy of *N&O.*

carnival was departing Danville, Virginia, to play the North Carolina State Fair. Funeral services were held in Raleigh the following day. Strates was laid to rest in Endicott, New York, on October 15 at 12:30 P.M. J. S. Dorton arranged a special tribute to Strates at the fairgrounds at that precise moment. Dorton had all the rides stopped and unloaded, all the concessions and shows shut down. A huge fireworks display began as a bugler played "Taps" and an honor guard from Fort Bragg lowered the flag in the grandstand area to half-mast. To end the observance, a helicopter hovering above the midway dropped thousands of red roses and chrysanthemums onto the crowd below. E. James Strates, his father's only son, had turned down law school to learn his father's business. Only three years out of Syracuse University at the time of his father's death, he took over the Strates operation, which he continues to manage. He attempted to maintain his father's basic approach to carnival management, but changing times eventually intervened.

Without question, from 1948 until the early 1970s, the girl shows ruled the Strates midway. Often Strates featured three girl shows, one of which was staffed by black performers, even during the segregation years. The featured girl show of the Strates midway of the 1950s, 1960s, and 1970s was the "Broadway to Hollywood Revue," operated by Jack and Bonnie Norman. Like the other Strates girl shows, it was essentially a burlesque show, modeled after the playbill at major burlesque houses in such venues as New York City and Atlantic City. It featured an emcee, a song-and-dance man who performed several numbers, a handful of musicians, and a few old burlesque comedy routines. But it was the strippers the male audience came to see, and the strippers were the undisputed stars of the show.

Some of the best-known strippers of the carnival world played the fair with the Strates shows. The story of the carnival burlesque show performers, including those who worked for the Strates shows, is recorded in *Girl Show: Into the Canvas World of Bump*

Girl shows ruled the midway from the 1930s to the early 1970s, with the Strates Shows always featuring at least three such shows during the era. Shown here is Siska, "Queen of Burlesque," at the 1954 state fair.
Photo courtesy of *N&O*.

Above, left: Performers in a girl-show front that played the Strates midway during the 1950s and early 1960s. Photo from Stencell.
Above, right: A mostly male audience responds to the "pitch" for Loreli's girl show at the 1955 state fair. Photo courtesy of *N&O*.

and Grind, by Al W. Stencell, himself a circus operator and now a historian of the carnival. Among the best known were Legs-A-Weigh Loreli, Rita Cortez, and Pagan Jones, who played the Strates shows in the 1950s and 1960s; Val Valentine and Bonnie Boyia, who appeared in the 1960s and 1970s; and Bambi Lane, who appeared at the fair during the 1980s. Jack Norman's Hollywood to Broadway Revue of 1970 typified the Strates girl show. In that year the Hollywood to Broadway Revue sent Culleene O'Day, Bambi Lane, Sheena the Savage, and Monique, the Blue Diamond of Burlesque, to the "bally," the small stage in front of the show used to entice crowds into the tent. There the girls gave a brief preview of the performance inside, while a barker "pitched" the audience, according to Stencell, telling them, among other things, "Don't worry about your friends seeing you go in the show . . . your friends are already there."

While the barkers promised risqué performances, the girl shows of the Strates midway for the most part offered legitimate burlesque show entertainment, which even by the 1950s was increasingly a quaint reflection of the sexual mores of a bygone Victorian era. There was no nudity, and much of the "stripping" in the strip acts was an illusion. As in 1895, the girls in the Strates shows of the last half of the twentieth century played under the watchful eyes of fair officials, in this case Wake County sheriff's deputies. In 1965, for example, deputy Wiley Jones made his rounds of the girl shows, telling operators: "No artifacts.

Rita Cortez employed a Latin theme in her show, which played on the Strates midway during the 1950s and 1960s. Photo from Stencell.

Behind the scenes, girl-show performers were neither glamorous nor sensual, just performers waiting for the next curtain call. Personnel of the Broadway to Hollywood Revue, which was the premier midway girl show with the Strates Shows from the 1950s to the 1970s. Photo from Stencell.

Keep the entire front of the body covered. Keep the bumps and grinds up, not down. If I get a beef, you've had it. And another thing, keep the hawking clean on the outside."

But the girls had to be able to entertain the crowds, which, especially through the 1970s, included large numbers of high school adolescents. Bonnie Boyia recalled in *Girl Show* that the Strates show was one of the toughest venues she played. "I followed a very young girl named Mary Lou Evans," she remembered. Mary Lou "wore blue cowboy boots, and little tight blue leather leotards under a cowboy vest and small skirt. She had a forty-inch bust and a twenty-inch waist, played the guitar and triple yodeled. Try topping that. I had to, as I closed the show."

Like all good strippers and exotic dancers, the girls of the Strates girl shows, Stencell notes, knew their overwhelmingly male audience. As Pagan Jones, a statuesque blonde burlesque star of the late 1950s and 1960s, put it, "The marks out there don't want a chorus

line; they don't want show tunes; they don't want choreography or ballet. All they want is sex—bumps and grinds—they want to see something they don't see at home." Nearly two decades later Val Valentine echoed those sentiments. "You have to be a ham," she told a *News and Observer* reporter in 1977. "It is a fantastic escape from reality for myself, for the crowds who come inside the tent to see what's going on in there. They don't come in here expecting to see girls who look like their wives and sisters. They want to see something glamorous and we try not to disappoint."

The sexual and social revolutions of the 1960s eventually doomed the girl show, but it died a slow death. According to E. James Strates, "With the integration problems that the country went through, the black show just ceased even to be able to pay for itself." As television and movies became increasingly sexually explicit, the girl show seemed tame indeed. The growth of topless bars and strip clubs in urban areas made the old burlesque

"*I followed a very young girl named Mary Lou Evans . . . [who] wore blue cowboy boots, and little tight blue leather leotards under a cowboy vest and small skirt. She had a forty-inch bust and a twenty-inch waist, played the guitar and triple yodeled. Try topping that. I had to, as I closed the show.*"

Left: Crystal Ames with a "talker" on the "bally" of a girl-show front at the 1968 fair. Photo courtesy of *N&O.*

Below: Once major midway attractions, freak shows, which featured numerous exhibits of natural "oddities" (some genuine, some fake), no longer appear at the state fair. Pictured here is a freak show at the 1967 fair. Photo courtesy of *N&O.*

acts seem pale in comparison, and show operators had to change their acts to survive the competition. The Normans, for example, dropped their novelty acts in 1965, placed more emphasis on strippers, and raised the ticket price. "We added strippers and we were selling hotter without jugglers and acrobats," said Jack Norman. "Listen, you can't have everything on the midway for kids, so we added sex without being offensive."

The Normans and the other girl shows were fighting a losing battle, however. In 1974 the Normans folded their tent, and the show of Gene "Broadway" Vaughan replaced their show until Vaughan's death in 1977, when various other producers played the Strates midway. In 1980 *News and Observer* feature writer Charles Craven observed that "In the old days the girlie shows were exciting and the comedians in them funny. They seem about shot now. The hard pornography of the cities has about taken over the long haired hay seed crowd." Craven was right, and the 1986 edition of the Strates shows was the last to feature burlesque under a canvas tent. As E. James Strates observed in 2000, "The country had gone way beyond anything that we ever had at the fair." Strates's observation was quite accurate. With

strip clubs and topless bars a part of the entertainment available in virtually any small American city, young women could make far more money working such local venues than they could with a carnival show. Ultimately, the shows folded not only because they were passé but also because they simply could not recruit performers.

The second-most-popular attractions on the midway, freak shows, like girl shows, enjoyed their heyday in the Strates carnival from the 1940s into the 1970s. Strates usually featured at least two freak shows and several individual attractions. Freak shows on the Strates midway varied little from those of the late nineteenth and early twentieth

Old-fashioned banners announce the attractions of this freak show to crowds at the 1968 fair. Photo courtesy of *N&O*.

centuries. Attractions included the usual fat man or fat woman, bearded ladies, strong men, flame eaters, and sword swallowers. One of Strates's more unusual attractions was William Parnell, victim of a rare skin disease and a native of Kenly, North Carolina, who appeared in the 1950s and 1960s billed as the "Alligator Man." Another star of the 1950s was the Viking Giant, billed as being eight feet eight inches tall, who amazed crowds by passing a silver dollar through his ring. An individual attraction that played as a standard feature of the Strates midway, as in most

screams, frantic movements, and such stunts as drinking "chicken blood" or "eating" mice. The staying power of the geek proved phenomenal, and in 2001 a Strates act included a highly sanitized version of the "geek"— a demented individual, who, the recorded pitch proclaimed, had been driven to his condition not by drink but by drugs. The Strates freak shows, as did many of its counterparts, also featured attractions that were actually illusions, such as the amazing headless woman, whose torso was inevitably connected to a contraption that kept her "alive."

> *The geek, originally a man supposedly driven to depravity by the abuse of strong drink, was always displayed as a warning against the evils of demon rum. He terrified his audience with horrid moans and screams, frantic movements, and such stunts as drinking "chicken blood" or "eating" mice.*

carnivals, was the "geek." The geek, originally a man supposedly driven to depravity by the abuse of strong drink, was always displayed as a warning against the evils of demon rum. He terrified his audience with horrid moans and

The popularity of freak shows suffered as social mores changed in the 1960s and 1970s. The notion that such exhibitions were demeaning to those being exhibited diminished the popularity of the shows—even though many of the "freaks" made an excellent living from displaying medical conditions that practically prohibited them from engaging in other kinds of work. E. James Strates recalled an incident that clearly illustrates the declining popularity of the freak shows. A family had taken a child with a physical handicap into a freak show, and the child became emotionally distraught. The family complained to E. James Strates that such exhibitions should be forbidden. That perception, Strates observed, gained increasing support among the general public, as indicated in 1990 by the passage by Congress of the Americans with Disabilities Act. But despite disapproval from a large segment of the public and the competition of the modern televised "freak show" (such as numerous talk shows featuring outrageous participants), the Strates freak shows never closed. The 2001 edition of the Strates carnival was the last to play the North Carolina State Fair.

Like the girl and freak shows, the Strates animal shows fell victim to changing attitudes. Into the 1960s, Strates offered at least one

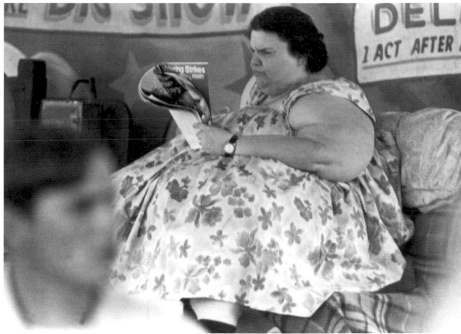

animal show, which usually featured animals considered to be dangerous to humans, such as bears, lions, and tigers. In addition, they almost always featured a snake handler, since the natural fear and loathing with which humans reacted to snakes essentially guaranteed an audience. Snake handlers often closed the show by biting a small green snake in two, an act that elicited oohs and ahs from the

The changing sensibilities of the American public made the exhibition of "freaks" on the fair midway, such as this Fat Lady at the 1968 state fair, less acceptable. Photo courtesy of *N&O*.

Animal shows have been a staple of the midway since the nineteenth century. Despite concerns that have surfaced in recent years about the well-being of animals in such shows, they continue to play the modern fair. Here tigers perform at an educational show about the demise of the world's Bengal tiger population at the 1999 fair. Photo courtesy of NCDA&CS.

horrified crowd. One day, E. James Strates recalled, as the handler bit the snake in half, someone in the audience cried out, "Oh, the poor snake." At that moment, Strates recalled, he knew that the time of the animal show was over.

The most important reason for the decline in appeal of the carnival show in the last half of the twentieth was not changing social and sexual mores, although they took their toll, but the rise in competing forms of entertainment and the increasing sophistication of the audience. With the spread of television into the homes of America, including those in rural North Carolina, even people in the state's most isolated areas had access to a constant stream of entertainment. As television programming became more sophisticated, and more sexually explicit, it proved increasingly difficult for carnival shows to meet the

competition it provided. Gradually, the carnival shows began to lose their preeminence on the midway to another attraction.

By the late 1970s, the rides had become the star attractions of the midway. They barely resembled the steam-powered merry-go-round and other rides that played the 1900 fair, or even the rides of the Sheesley shows of the 1920s, which had been powered by gasoline engines. A powerful diesel engine propelled the modern rides, which each year became bigger, taller, and faster. Not even the most traditional of the midway rides, such as the Ferris wheel, resisted the impulse toward the spectacular. By the late 1950s, the Strates shows included a gigantic double Ferris wheel, stacked one atop the other, which allowed the entire frame, as well as each

Not quite a ride, yet not a show, the Fun House was nevertheless a featured attraction on the midway of the twentieth-century fair. This brightly lit version played the 1984 fair. Photo courtesy of NCDA&CS.

While rides at the fair date back to the 1890s, they came to dominate the midway only toward the end of the twentieth century. Powered by enormous diesel engines and ablaze with computerized displays of colored lights, modern rides offer fairgoers instantaneous thrills and an accompanying adrenaline rush. Shown here is the "TopSpin" at the 1997 fair. Photo courtesy of NCDA&CS.

individual wheel, to rotate. In 1980 Strates introduced the "giant" Ferris wheel, which was the tallest structure on the fairgrounds. In that year the Strates shows featured eighty different rides, some of which, like the "Tidal Wave," a huge structure shaped like a ship that rocked from side to side, reflected the trend to more spectacular machines. By the 1990s, the Strates shows were importing even larger, more complex rides, primarily from European manufacturers, some with price tags in excess of one million dollars. A typical example was the "TopSpin," a ride that carried forty-two people, which E. James Strates described as "a beautiful ride, a production in itself." Many of the new rides were so large and heavy that they were carried about the nation on rail cars. Their size increased their economic efficiency, allowing far more passengers to board for each thrill-packed experience.

The "Himalaya" at the 1984 fair. Photo courtesy of NCDA&CS.

The "Thunder Bolt" at the 1984 fair. Photo courtesy of NCDA&CS.

Riders wait their turn to be rocked and rolled on a "theme" ride at the 2002 fair. Photo courtesy of the author.

For the more timid fairgoer, traditional rides provide a less harrowing, more pleasant experience, one that all members of the family can enjoy. Pictured here is a carousel at the 1984 fair. Photo courtesy of NCDA&CS.

A Ferris wheel at night at the 1995 fair. Photo courtesy of NCDA&CS.

A Ferris wheel at night at the 1995 fair. Photo courtesy of NCDA&CS.

A giant swing at the 1996 fair. Photo courtesy of NCDA&CS.

The twentieth-century fair always offered "kiddie rides" for the pleasure of the very young and their parents. Here children ride a toy tank at the 1954 fair. Photo courtesy of *N&O*.

A growing diversity in the types of rides allowed them to appeal to a wider range of individual tastes. Adorned with flashing neon lights, the huge machines presented fairgoers with a dizzying, eye-popping, computer-controlled display of raucous, undulating, pulsing motion, which illuminated the night sky and was visible for miles. The rides, with the thrills and spectacular displays they delivered, became the main attraction on the midway.

In many ways, the games of the midway carnival remained the attraction least influenced by change in the twentieth century. Just as the case at the nineteenth-century fairs, modern operators of various and sundry "games of skill" sought to entice fairgoers with opportunities to display their skills by winning any of a variety of "prizes." Indeed, photographs of a child or girlfriend hugging a stuffed animal won by a father or boyfriend became emblematic of the post–World War II fair. Fairgoers, overwhelmingly male, attempted to knock over stacked "milk" bottles or weighted cat-like figurines with baseballs, to throw softballs into tilted bushel baskets or upright milk cans, to shoot basketballs through a standard netted hoop, or to

obliterate a design on a paper target with well-placed rifle shots. Primarily younger players of both sexes tossed pennies, then dimes and quarters, at dishes positioned on a platform, hoping to win the dish upon which they could get a coin to rest. They tried to drop a ring tied to a pole around the neck of a pop bottle, threw darts at inflated balloons, or squeezed the trigger on a water gun to inflate a device

With gambling prohibited at the state fair in 1891, fairgoers turned to "games of skill" on the midway to pay for the opportunity to win a variety of prizes. One of the traditional games involves pitching softballs into a milk can, shown here at the 1983 fair. Photo courtesy of NCDA&CS.

The stuffed animal has remained the most coveted prize at the fair's games of chance throughout the twentieth century. Here a young girl hugs a stuffed bear at the 1946 fair. Photo courtesy of N&O.

Above: Stuffed animals waiting to be won at a midway game at the 2002 fair. Photo courtesy of the author.

Right: All the "games of skill" on the fair's midway actually demand some level of performance on the part of the player, not mere luck. These young men used their basketball-shooting skills to collect a number of stuffed animal trophies at the 2002 fair. Photo courtesy of the author.

Below: Some midway games required less skill of participants than did others. The rate at which these players at the 1996 fair could fill a target with a water gun determined who won the always-desirable stuffed animal prize. Photo courtesy of NCDA&CS.

faster than their competitors. Failure invariably drew the operators' encouragement to try again.

As in the nineteenth century, policing the game operators proved a never ending task. In the early twentieth century, the North Carolina State Agricultural Society employed a private, deputized force to police the fair, including all game operators. When the state acquired the fair in 1928, that job fell to the deputies of the Wake County Sheriff's Department. Enforcement centered on two concerns. First, state law prohibited gambling, so all games had to have some element of "skill." Second, game operators could, and sometimes did, engage in shady practices, such as rigging games so that they could not be won or confusing customers about the amount of money they owed for playing. In 1937, when the fair first opened as a division within the Department of Agriculture, J. S. Dorton assured the public that no "gyp artists" would be permitted on its midway, and deputies did their best to see that his promise was kept.

In 1948 Dorton brought the Strates carnival to the fair in part because of its reputation for "good, clean shows." Almost twenty years earlier, in 1929, four years before Hamid had his run-in with corrupt midway "gypsies," Wake County deputy Wiley Jones began his efforts to keep the "carneys" straight. In 1965, his last year on the job as the "unofficial police chief" of the fair, he checked tents to ensure that they held no illegal slot machines and gave carnival game operators the same advice he had dispensed more than a quarter-century earlier: "Give the customer his change before he throws the first ball. Don't allow him to leave his change on the counter and no 'buildups' (additional throws for unpocketed change). No game over fifty cents, you have to post the price in plain sight." In 1983 Wake County sheriff John Baker, after tossing a few footballs at a moving circle in a game operator's tent, admitted that his department was ill equipped to police the games because such games "could check out OK and be rigged in fifteen minutes." But he also commented that "We have had very few complaints from fairgoers. I am very pleased with the James E. Strates operation." Tom Nemia, the manager of the Strates game operators, confirmed, "Every game here can be won. They are not controlled by the operators." In general, Dorton's

belief that the Strates operation would run "clean," or fair, games proved correct. Modern fairgoers realize that their chances of winning are slim but not impossible. And the chance for some stalwart male to win, at an average cost well beyond its purchase price, a stuffed bear, gorilla, or giraffe to impress his wife, children, or girlfriend, or for a kid, after untold tries to make a coin remain in a dish worth no more than 25 cents, to "win" the dish, remained a hallowed fair tradition.

Especially in the twentieth century, food has contributed significantly to creating the state fair experience. Indeed, for many fairgoers it is an indispensable part of the carnival atmosphere. Although some food vendors appeared at the fairs of the nineteenth century, the food items most North Carolinians

associate with the carnival did not appear until the twentieth century and did not become staples for midway and other food vendors until the 1920s. Cotton candy, perhaps the quintessential carnival delicacy, provides a perfect example of the process by which a particular food became associated with carnivals, circuses, and fairs. While the origin of the incredibly sticky confection made of colored sugar spun in a centrifuge is in dispute, it was most likely introduced at the Ringling Brothers Circus in 1900 and quickly spread to other carnival-like attractions. When the state fair first opened at its current location in 1928, cotton candy had already become a traditional midway treat.

Other food items associated with the midway share a similar history. Hot dogs, a variant of German sausages, descended from the frankfurter, which was being sold in a roll in the St. Louis German community by the 1880s. Hot dogs are named, however, after New York newspaper cartoons of the early twentieth century depicting a dachshund in a bun and implying that the "hot dogs" sold at Coney Island and other area resorts contained

Above: Children flock to a cotton candy concession at the 1955 fair. Photo courtesy of *N&O.*

Below: Sharing a hot dog with a good friend at the state fair could be an enjoyable, if somewhat awkward and messy, experience. Here two girls share a hot dog at the 1955 fair. Photo courtesy of *N&O.*

dog meat. Heavily promoted by Coney Island vendors, the most famous of whom was Nathan Handwerker, the hot dog became a national obsession by the 1920s. The hamburger, descended from the hamburger steak made famous by the 1904 St. Louis Exposition, had evolved into a grilled ground-beef patty served on a bun prior to World War I and, like the hot dog, was a national phenomenon by the 1920s. After World War I the rise of major carnival operations, which traveled long distances via rail or motor truck, rapidly introduced carnival foods, of which the most important were cotton candy, candy apples, hot dogs, hamburgers, and French-fried potatoes, to fair audiences throughout the nation. The major carnival organizations that played the twentieth-century North Carolina State Fair, including the Strates shows, included food vendors who plied their delicacies to hungry fair patrons.

Feeding fair crowds, of course, presented an economic opportunity long before the development of the modern midway, and well before the end of the nineteenth century civic groups and churches in the Raleigh area took advantage of it. The 1884 fair featured food booths at which such groups offered meals to hungry fairgoers. By the turn of the twentieth

century, booths run by local churches and organizations such as St. Mary's Guild had become a staple of the fair experience. With the construction of the current fairgrounds in 1928, the hastily assembled food stands of carnival vendors faced even more competition for the fairgoers' dollar from a variety of food booths operated by North Carolinians. Recognizing an excellent fund-raising opportunity, churches, fraternal organizations, civic clubs, and other groups primarily from Wake and surrounding counties began to contract with the fair's management for increased booth space. By the 1930s, these organizations hawked their menu items from a row of ramshackle, dirt- or sawdust-floored wooden structures near the grandstands. A fire destroyed many of those structures in 1963, but some survived, were refurbished, given concrete floors and modern electrical circuitry, and continue to serve fair patrons. Such concessions, most of which are operated by organizations whose members volunteer their labor, offer the modern fairgoer a diet designed to appall a cardiologist. Barbecue, both pork and chicken, are favorites, as are baked beans, cut green beans, and small Irish potatoes, each invariably flavored with bacon or other pork cuts; hush puppies, a deep-fried

Since the fair has been at its present location, regional nonprofit organizations have operated food booths there as a means of raising funds. This civic organization boasted the world's greatest ham biscuits at the 2002 fair. Photo courtesy of the author.

corn bread; French fries; and mayonnaise-based cole slaw. Competition among the booths run by local organizations adds to the fair's festive ambience. The competition for the best ham biscuits among church groups sponsoring food booths, for example, is fierce but friendly, and fairgoing biscuit eaters are actually the winners.

By 1965 more than a hundred food stands and booths served fairgoers, and the amounts of food they offered was amazing. Since vendors purchased from different suppliers, figures on total food consumption at a given fair are hard to obtain; but records of one company from the 1964 fair provide an insight into how important food had become to the fair. In that year a single major wholesale meat distributor in the Raleigh area, Jesse Jones Sausage, delivered thirty-five thousand pounds of hamburgers, hot dogs, sausage patties, chili, and bacon to the fair's food vendors.

During the last quarter of the twentieth century, a variety of foods more familiar in the eastern and Midwestern United States invaded the fair, but few have lowered the fat content of the fairgoer's diet. The newer items include funnel cakes, a fried confection that originated as a Pennsylvania Dutch breakfast disk; elephant ears, deep-fried, sugared slabs of dough similar to a doughnut; Polish and Italian sausages, grilled with onions and peppers and lots of grease; deep-fried onion rings and whole onions; and "desserts" such as frozen bananas dipped in chocolate, which compete with the more traditional block of vanilla ice cream dipped in chocolate, covered with chopped nuts, and served on a cone.

Modern fairgoers, perhaps because of the emphasis health professionals place upon a "healthy" low-fat diet, continue to enjoy the

Some food concessions have become traditions in their own right. Shown here is the Al's French Fries concession, a perennial favorite with fairgoers, at the 1984 fair. Photo courtesy of NCDA&CS.

Above: Funnel cakes, served at this fair booth in 2002, are relatively new additions to the state fair menu. Photo courtesy of the author.

Right: The Charlie Barefoot & Sons booth (here shown at the 2002 fair) boasts of a fifty-four-year history of serving fairgoers. Photo courtesy of the author.

Above: Many of those who attend the fair annually gravitate toward their favorite food booths. Shown here is "restaurant row" at the 2002 fair. Photo courtesy of the author.

Below: In 2002, for the first time in fifty-two years, a new carnival, Amusements of America, appeared at the state fair. Shown here is the new carnival's center of operations on the midway. Photo courtesy of the author.

guilty pleasures of indulging on the delicious but decidedly "unhealthy" fare offered both at the stands of carnival food vendors and at the booths of community organizations. Indeed, if surveys of fairgoers are correct, food is the major reason most North Carolinians come to the fair. In a 1996 survey, for example, fairgoers were asked to place a check mark beside each item in a list of attractions offered by the fair. An astounding 70 percent checked food. (The second-most-frequent response, agricultural exhibits, received checks from only 54 percent of those surveyed.) The survey's findings concerning food are less dramatic, however, when all possible responses are considered. Forty-seven percent also selected

livestock exhibits, and an equal 47 percent checked the "Village of Yesteryear," both of which are agricultural displays. Even so, it is obvious that modern fairgoers, like their counterparts of the early twentieth century, come to the state fair prepared to enjoy a culinary experience, if not an Epicurean one.

Just as the mix of foods, rides, shows, concessions, and games continues to change on the carnival midway, so does the carnival operator at the state fair. The 2001 edition of the fair saw the Strates shows' last stand on the midway. Following her election as commissioner of agriculture in 2000, Meg Scott Phipps requested bids from carnival operators nationwide. Although the Florida-based Strates shows entered a bid along with ten other large carnivals, in a controversial decision the award went to Amusements of America, a New Jersey-based company and one of the largest in America. Supporters of the decision cited the fact that the carnival was never put up for bid under Commissioner Jim Graham and that Strates paid slightly less for the right to operate the carnival than was customary in the business. Supporters also said that the fair needed a "fresh" look on the midway. Opponents countered that the Strates shows were exceptionally responsible and reputable, that the familiarity of the state fair staff with the Strates organization made for smoother operations, and that the Strates shows spent

large sums of money with North Carolina vendors. There was also considerable sentimental support for Strates, especially among some who saw the change as disparaging of the Jim Graham era. The controversy ultimately resulted in a challenge to the commissioner's authority to make the change, but the challenge was refuted in a ruling by the state attorney general.

In October 2002, for the first time in fifty-three years, a carnival other than the Strates shows played the North Carolina State Fair. On the surface, little seemed to have changed, except that the carnival arrived by truck, not by train. Fairgoers still maneuvered through fun-house obstacle courses, sought to beat the odds and perhaps win a stuffed toy in a "game of skill," purchased hot dogs and candied apples from carnival vendors, paid their money to see exhibits of the weird and the wondrous, and lined up to purchase tickets to ride on machines that would scare them half to death. The carnival continued to deliver what it had always delivered—a brief respite from the work-a-day world, a few moments of escape into fantasy and frivolity, and memories of happy times. But the circumstances surrounding the end of the Strates era would prove to have enormous consequences for the state fair and for Commissioner Phipps.

And a Good Time Was Had by All

The Fair as Social Institution

The Fair's Social Appeal as a Founding Concept

Few events better display the diversity, vitality, and essential optimism of the people of North Carolina than the ten-day-long exhibition and extravaganza known as the North Carolina State Fair. The high esteem in which North Carolinians hold the fair is anything but new; the fair has been one of the state's premier social institutions since its inception. During the nineteenth century, fair week was the social event of the year, an occasion anticipated by all elements of society. Fair week afforded North Carolinians a time for participating in meetings of professional, civic, and fraternal organizations; attending galas, balls, and other social activities; renewing friendships with people from other sections of the state; taking in the sights of the capital city; gambling; drinking; and having a good time in general. It offered fairgoers a week of bands, parades, balls, tournaments, and political orations. From city and farm, from every walk of life and all social stations, the people of North Carolina converged on Raleigh during fair week, bringing with them the desire for a respite from their daily labors and the willingness to expend money to support

---

Enjoying a wide variety of fair foods with family and friends is the activity fairgoers most anticipate. Crowds stream through a 2002 state fair thoroughfare lined with food concessions. Photo courtesy of the author.

their activities, both at the fairgrounds and in the city. Within less than a decade after its founding, the fair had become a social institution the people of North Carolina could hardly do without.

The founders of the North Carolina State Agricultural Society recognized, indeed, counted upon, the social appeal of the week-long fair to attract crowds. With an eye toward encouraging attendance, the society's leaders selected Raleigh, the capital city, as the fair site and a late October date for the fair—a time

destinies of the state. Three of the society's postwar presidents are particularly exemplary in that regard. Bennehan Cameron, the only surviving son of Paul C. Cameron, inherited his father's wealth and invested heavily in railroads and textiles; Thomas M. Holt, son of the state's leading antebellum textile mill owner, Edwin M. Holt, and nephew of William R. Holt, developed his own textile firm, served as president of the North Carolina Railroad, and in 1890 was elected governor of the state. Kemp P. Battle, scion of one of the state's most

> *W*ith an eye toward encouraging attendance, the society's leaders selected Raleigh, the capital city, as the fair site and a late October date for the fair—a time when most farmers had completed their harvests.

when most farmers had completed their harvests. Leaders of the society believed that the fair could inform, "while adding also more interest, excitement and pleasure to the [agricultural] calling." They also saw the fair as a means of developing pride in the state and of "breaking down the barriers of sectional prejudices that divide our citizens [and] bringing them together in pleasant association." Largely because of its social appeal, the society reestablished the fair in 1869 after the trauma of the Civil War had forced its closure in 1861. Although the revived fair's instructional activities expanded constantly, its entertainment value and social appeal helped make it the unqualified success it became throughout the remainder of the nineteenth century.

A major reason for the sustained success of the fair as a social institution was the prominence of the leadership of the Agricultural Society. Throughout the nineteenth century, members of the state's most prominent families held offices within the society and lent their prestige to the fair. In the antebellum era, in addition to Thomas Ruffin, Kenneth Rayner, and John Dancy, other well-known and respected figures gave the society and its fair their unqualified support. Among them were Paul C. Cameron, Weldon N. Edwards, and William R. Holt, each members of one of the state's most powerful families.

The society's leadership after the Civil War was no less impressive; indeed, it was representative of the leadership class that controlled the economic and political

noted families, was a prominent lawyer and judge and from 1876 to 1891 served as president of the University of North Carolina. The support of such notable North Carolina families as the Camerons, the Holts, the Ruffins, and the Battles clearly enhanced the fair's appeal.

Social Activities at the Nineteenth-Century Fair

Another reason why the fair prospered was simply that it put on a good show. It thrived on pomp and pageantry, and few attractions surpassed bands in lending a large measure of both to the occasion. The Agricultural Society hired bands to attract and amuse crowds at the inaugural fair of 1853. A rousing success, bands quickly became a fixture on the fair's program. Participating bands usually represented their hometowns and thus embodied a spirit of civic pride and competition, as well as exuberance. Among the many bands to perform at antebellum fairs, one from Salisbury appears to have been the favorite. In the postwar era, as crowds grew larger and more bands were required to entertain them, the Salisbury band remained a favorite, but bands from throughout the state rendered a variety of martial and popular tunes for fairgoers. Among the towns often represented were Salem, Concord, Carthage, Winston, and Raleigh. In 1886, for example, three bands performed, and in 1880 at least two appeared.

Over the years a remarkable variety of bands played the fair, indicating that North Carolinians had rather eclectic musical tastes—or, on the other hand, were so musically unsophisticated as to be nondiscriminatory. String bands, brass bands, marching bands, military bands, and silver cornet bands all played for and pleased fair crowds.

Military and paramilitary drill and marching units provided antebellum fairs with an additional dash of pageantry. Such organizations enjoyed widespread popularity among the white population of North Carolina and the rest of the South for a variety of reasons. The American South was a patriarchal society, in which men followed a strict, and well understood, code of personal honor. Physical courage, especially when expressed in military exploits, was highly valued. From colonial days, men who would be leaders had been expected to display their courage by serving in military organizations. But there was another reason, far less gallant, for the region's emphasis on martial organizations. They were, in fact, absolutely essential for the maintenance of order in a slave society. Put bluntly, North Carolina, like other southern states, required militia units and slave patrols to ensure that its slave population remained under control. Thus, service in such units reflected both notions of honor and necessity, and public displays by military organizations reassured white citizens that they resided in safety with their human property.

Like bands, drill teams often represented towns; others came from military schools. In 1855 two Raleigh units, the Oak City Guards and the Independent Guards, performed at the fair, as did cadets from Lovejoy Academy, Raleigh's most fashionable boys' school. Cadets from Hillsborough Military Academy "astonished" crowds with their performance of a zouave drill at the 1860 fair, according to one of their own. Although he found the fair unimpressive, the cadet, true to the region's code of chivalry, extolled the charms of "the large number of young ladies from [Raleigh's] St. Mary's School" that he encountered at the fairgrounds.

Even though the outcome of the Civil War brought to most southerners the realization that they had sacrificed many lives and considerable material treasure to a "Lost Cause," it did not significantly alter the traditional southern appreciation of the martial spirit. The terrible loss of life sustained by North Carolina soldiers in that conflict (19,673 North Carolina soldiers fell on the field of battle—a full quarter of all Confederate battlefield deaths) had no long-term effect upon the esteem in which North Carolinians held the warrior. Indeed, like most other southerners, North Carolinians reveled in celebrating the gallantry and bravery of the Confederate soldier. Drill teams provided an ideal means of celebrating that spirit, and practically every military school and town in the state that boasted a drill unit sent it to the fair. In 1876,

Parading military units were extremely popular with pre-Civil War crowds and remained so even after the war. Shown here are the LaGrange Cadets at the 1884 State Exposition.
Photo courtesy of A&H.

for example, two groups from Raleigh, one each from Wilmington, New Bern, and Fayetteville, and the cadets of Bingham Academy in Orange County performed there. The following year governors Zebulon Vance of North Carolina and Wade Hampton of South Carolina reviewed twenty military units at the fairgrounds. During the late nineteenth century, units from neighboring states performed at the fair. In 1883 Virginia units augmented those from North Carolina, and in 1893 Maryland sent a contingent of martial performers

to the festivities. By the end of the century, the drills had escalated to mock battles, such as one staged at the 1896 fair by the Pitt Rifles and the Franklin Guards.

North Carolinians obviously saw the martial arts as an almost sacred calling, and groups that parodied the military proved unsuccessful as fair attractions. The reception granted the Royal Raleigh Ringtail Rousers, which paraded in rags and pumpkin heads at the 1858 fair, offers a case in point. A Raleigh reporter described the Rousers as "a crowd of

From Reconstruction to the rise of racial segregation at the turn of the twentieth century, "colored" military units remained organized in many North Carolina communities, though they were rarely featured at state fairs. The Colored Military Company of Raleigh performed at the 1884 State Exposition. Photo courtesy of A&H.

grotesque beings . . . unworthy of mention," whose performance could be enjoyed only "by negroes and children." The group's initial appearance, as the reporter hoped, proved its last. The 1875 fair ended with a dress parade of the Mulligan Guards and the Grand Panjan drums of the Noble Order of Flapdoodles. They, too, appeared but once. Such seemingly comical displays would seem sure attractions; their failure indicates that nineteenth-century North Carolinians were too emotionally involved with the martial arts to view them with a sense of humor.

Parades, with all their spectacle and excitement, rivaled bands and military drill units in popularity. During the antebellum era the fair marshals and officers of the North Carolina State Agricultural Society, all mounted on horseback, led a procession to the fairgrounds to open the fair. In the postwar period, opening-day processions were transformed into elaborate parades. Bands, bicyclists, marching units, and clowns lent the parades color and excitement, but the processions also included promotional floats sponsored by various industrial exhibitors at the fair. In 1880, for

example, Durham's W. T. Blackwell Tobacco Company, then one of the state's leading tobacco firms, entered a float on which African American employees manufactured chewing tobacco; and the float of Edwards and Broughton Company of Raleigh carried an operating printing press. Other firms sponsored floats depicting the manufacture of leather goods, boilers, and granite monuments. The parade at the 1892 fair, held in conjunction with the Raleigh centennial celebration, was the most spectacular of the century. Greater than two miles in length, it was composed of more than two hundred mounted fair marshals, forty floats, numerous government officials, and several marching bands and military units.

The annual address, invariably delivered by a prominent North Carolinian, became another of the fair's major social attractions. The oration, which originated at the inaugural fair as an instructional lecture on scientific agriculture, became essentially a social and political event by the end of the antebellum era. After the Civil War, the governor normally delivered the address, and lesser politicians,

This drummer was part of the Colored Military Company of Fayetteville at the 1884 State Exposition, also pictured above. Photo courtesy of A&H.

dignitaries, and socially prominent North Carolinians
made it a point to be in
attendance. On occa-
sion, other powerful
political figures
delivered the
address, as was
the case in 1893
when United
States senator
Zebulon
Vance de-
nounced the
Populist Party
and appealed
to North Car-
olina farmers to
remain loyal to
the Democratic
cause. Indeed, by
the end of the nine-
teenth century the fair
had become a prominent
venue for politicians seeking
statewide office. The Agricultural
Society occasionally received requests to
make politics the special attraction, as was the
case in 1896 when North Carolina Democrats
exerted their efforts, without success, to obtain as a featured speaker William Jennings

Bryan, the party's presidential nominee of
that year.

Among the many sporting events that appealed to fair crowds, horse racing was the
most popular. Horse racing had a long and
honorable history in North Carolina, as it did
in other areas of the South. While presenting
quite a spectacle, it also celebrated the role of
the horse in society, provided an excellent opportunity for wagering among friends, and
rewarded spectators with an exciting contest.
The original fairgrounds, located east of the
Capitol, had a racetrack near its center. The
society constructed a new half-mile track
when the fair moved to the western edge of
the city in 1873. Although purses were small,
throughout the 1870s and 1880s the fair attracted some of the finest horses from across
the state and from several eastern states. In
1882 horses from several North Carolina
counties, as well as from New York, Virginia,
and South Carolina, went to the starting gates.

By the early 1890s, however, the fair was ex-
periencing difficulty in attracting quality
horses, perhaps because of the depressed national economy. Although public interest in
the sport remained high, the Agricultural Society acknowledged that poor-quality horses
had led to a decline in racetrack crowds. Since
the society charged admission to the races,
the smaller crowds translated into decreased

**Above: Appearances
by political leaders
became a common
occurrence at the state
fair during the late
nineteenth century.
In 1893 the state's
most revered political
leader, former gover-
nor and United States
senator Zebulon B.
Vance, addressed fair
crowds.** Photo courtesy
of A&H.

**Right: Racing was
enormously popular
with fairgoers of the
nineteenth century.
This engraving from
the Raleigh *Daily State
Chronicle,* titled "The
Finish," shows a sulky
driver crossing the
finish line during a
race at the 1890 state
fair.**

THE FINISH.

revenue. The need to improve the racing program prompted the society to elect Bennehan Cameron president in 1895, in a large measure because he was known as an accomplished horseman who was well acquainted with other horsemen of the East Coast. Cameron bred horses and cattle at his Stagville plantation in Durham County. Upon his election, Cameron set out to reestablish an excellent racing program. In a survey of horsemen throughout the southeastern United States, he ascertained that many of them considered North Carolina tracks inferior, the judges poor, and the purses small—all reasons for keeping their horses out of the state. The horsemen suggested a number of reforms for the fair's program, including the hiring of more competent judges and starters, better track crews, a more equitable balance between trotting and running races, and larger purses.

Cameron adopted most of those reforms for the 1896 races and added some touches of his own. To appeal to the average owner, he inaugurated races for "gentlemen's road horses," as well as a race for jumpers. With thoroughbreds, trotters, jumpers, and what was essentially an "open" class, Cameron's racing program appealed to all social classes, from the large stable owner to the farmer with a fast "road horse." The revised program proved popular, attracting enough spectators through the track gates to provide the society with revenues with which to bolster its then shaky financial standing.

Although no other spectator sports challenged the primacy of horse racing, a number of them vied for the attention of post-Civil War fairgoers. Cherokee Indians played their native game of handball, which resembled lacrosse, at the 1871 fair; they were well received and were invited to subsequent fairs. The fair added baseball to its list of attractions in 1873 when teams from Pittsboro and Goldsboro played each other. The "American

Bennehan Cameron, a noted horseman, was among the state's wealthiest citizens of the late nineteenth and early twentieth centuries. He inherited his father's plantations and invested extensively in banking and railroads. Photo courtesy of A&H.

Below: From the inaugural fair it sponsored in 1853 until the fair became a state agency, the North Carolina State Agricultural Society relied heavily upon the wealth, prestige, and social prominence of its members to ensure its ability to stage the fair. This certificate of life membership in the society, dated 1873, belonged to D. S. Waits of Raleigh. Image courtesy of Blankinship.

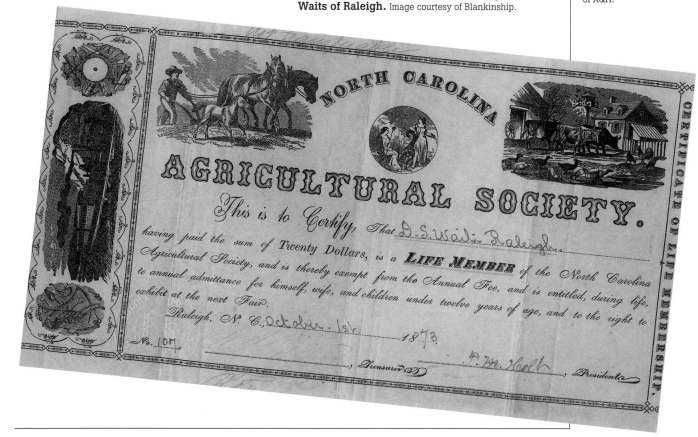

The North Carolina College of Agriculture and Mechanical Arts (known as the North Carolina College of Agriculture and Engineering after 1917) football team played at the state fairs of the late nineteenth and early twentieth centuries. The fairgrounds were immediately across Hillsborough Street from the college from 1873 to 1925. Photo courtesy of A&H.

pastime" soon became a regular feature. Baseball was extremely popular in the 1870s, but its appeal declined in the late 1880s, perhaps because by that time most communities had one or more teams. In 1892 students from the North Carolina College of Agriculture and Mechanical Arts introduced the more physical sport of football to fair crowds. Afterward, a number of other football contests were scheduled at the fair, and the contests became something of a fixture in the schedule of fair-week events.

While spectator sports attracted the larger crowds, sporting events in which the "average" fairgoer could participate likewise became major attractions. Pigeon-shooting contests, which celebrated the sharpshooting ability of a people who loved their guns almost as much as their horses, enjoyed great popularity. To add to the appeal of the contests, fair officials

Agricultural Society added bicycle races, archery contests, and broad-saber tournaments to the program of the 1881 fair. The saber tournaments, in which young men dressed as knights went at one another with fake swords, were taken directly from the pages of Sir Walter Scott's novels, and complemented the prevailing romanticized notions about the bravery of Confederate soldiers, especially Confederate officers. The tournaments were also a splendid example of the southerner's continuing devotion to pre-Civil War concepts of romantic chivalry, and they gained enormous popularity in the 1890s. True to the chivalric ideal, the winner received the honor of crowning the Queen of Love and Beauty at an annual Coronation Ball, usually held at the end of fair week.

Since the post-Civil War fair attracted such large crowds, a variety of organizations, seek-

While spectator sports attracted the larger crowds, sporting events in which the "average" fairgoer could participate likewise became major attractions. Pigeon-shooting contests, which celebrated the sharpshooting ability of a people who loved their guns almost as much as their horses, enjoyed great popularity.

introduced additional shooting targets, including glass balls and Irish potatoes. Foot races, inaugurated in 1873, remained popular throughout the years following the Civil War. In an effort to attract more young people, the

ing to improve attendance, began to hold their meetings either at the fair or in Raleigh during fair week. By so doing, they substantially increased the fair's social significance. North Carolina's Mexican War veterans convened at the 1873 fair, and the Grangers, the largest organization of farmers in the nation at the time, assembled at the fair of 1877. In 1881 the state's Confederate veterans held their annual reunion at the fair, an event that became a permanent feature of fair week. Another group to make the fair its permanent convention headquarters was the Northern Settlers Association of North Carolina. That organization, composed of northerners who had adopted North Carolina as their home, held its first convention, attended by more than one thousand people, in 1886. The 1887 fair hosted a reunion of expatriate Tar Heels. Either the gathering reinforced the expatriates' decision to depart their native state or caused them to stay home, for the group did not appear at subsequent fairs.

In an effort to increase attendance, the Agricultural Society had begun to promote "special" fair days in the 1880s. School day, begun in 1880, became an annual feature and the most popular of them. School day featured pig races, bag races, wheelbarrow races, greased-pole climbing, and other events designed to appeal to young people. In 1891 the society established a special "colored day" for the state's black citizens, a response to increased segregationist sentiments among the white population. Prior to the 1890s the fair was integrated, even though several leaders of

Raleigh's black community established their own fair in 1879. As mentioned previously, the "colored fair," sponsored by the North Carolina Industrial Association, sought to display the educational, industrial, and agricultural accomplishments of the state's black citizenry. Yet, the Negro fair did not discourage black attendance at the state fair. In 1882, for example, more blacks attended the state fair than attended their own industrial fair, held a week earlier. Even after the designation of "colored day" at the state fair, blacks continued to attend in considerable numbers each day of fair week, as they had since the fair began. The rigid segregation of the fair, and of all aspects of life in North Carolina, came only after the success of white supremacists in capturing control of the state's political machinery in the bitter, racially charged campaigns of 1900. The racial policies of the white supremacy Democrats, which accurately reflected the racial sentiments of the vast majority of white North Carolinians, locked the state, and the fair, into a policy of rigid racial segregation for two-thirds of the twentieth century.

Despite its availability, alcohol caused fair officials far fewer problems than did gambling. From the fair's inception, the society banned alcoholic beverages on the fairgrounds, and during the antebellum period drunkenness seems not to have been a

Picnics have always been a popular activity at state fairs. Here a group of young women enjoy lunch on the fairgrounds during the 1955 fair.
Photo courtesy of *N&O*.

PASSING THE TIME AWAY AT THE FAIR

SOME THINGS THAT ARE GOING ON AT THE FAIR.

Although alcoholic beverages were not available at the fairgrounds, they were legal and readily available to nineteenth-century fairgoers. Alcohol was not a problem at the state fair, but pickpockets were. A cartoon from the Raleigh *North Carolinian* showing pickpockets at work at the 1895 fair.

problem. But with the growth in size of fair crowds in the post-Civil War era, the situation changed. The fair marshals managed, nevertheless, to direct thirsty visitors to Raleigh's taverns. While that policy pleased tavern keepers and kept the fairgrounds free of drunkards, it also created a problem for the city, in which the taverns did a profitable business. As one observer of the scene noted, during fair week Raleigh saloons "were three deep at night with frantic calls for the beverage and for a wonder there was but little drunkenness."

With large crowds in attendance and alcoholic beverages so readily available, the fair ultimately encountered some difficulties with those who could not limit their intake. Such incidents, however, were usually minor. In 1881, for example, the press complained of too much drinking outside the fair's gates and requested that better order be kept on the grounds. Throughout the nineteenth century, the bars and saloons of Raleigh remained open twenty-four hours a day for the duration of the fair; yet, with the exception of some minor disturbances, the fair experienced surprisingly little trouble from persons under the influence of alcohol. No serious incident appears to have stemmed from the behavior of drunken fairgoers, which may indicate only that inebriated North Carolinians were happy in that condition.

Considerably enhancing the appeal of the nineteenth-century state fair was the fact that the fair was not just a singular event. Rather, from the antebellum era forward, it evolved into an integral part of Raleigh's overall social

life. The process of blending the fair into regular events in the social life of the capital city continued at an accelerated pace in the post-Civil War years. That process was natural, given the era's slow means of transportation, which encouraged many antebellum visitors to obtain accommodations in Raleigh and to remain in the city for the duration of the fair. The state's railroads made it easier for North Carolinians to travel to Raleigh for an extended stay by offering special excursions to the fair, by reducing rail fares for fairgoers, and by heavily advertising both policies. The railroads' promotional activities added passengers while at the same time increasing the size of the crowds that poured into the city for fair week. At the first fair in 1853, for example, Raleigh was overrun by fairgoers, who made up the largest crowd the city had seen "since the great Whig Convention in 1840."

Each year, a parade of North Carolina's most prominent citizens went to Raleigh in October to attend both the fair and the capital city's social events. The list of those attending included well-known politicians, educators, and agricultural leaders, as well as the state's social elite. In the 1860s, for example, honored guests at the fair included Kenneth Rayner; Kemp P. Battle; Thomas Ruffin; Edmund Ruffin, the noted Virginia agricultural reformer and editor; former United States senator Bedford Brown of Caswell County; and Daniel M. Barringer, former Whig congressman from Cabarrus County. Naturally, politicians took advantage of the large and influential audience the fair provided. In the antebellum era especially, politicians spoke at the Wake County Courthouse each night of the fair. Inasmuch as the orations of many political figures represented amusing entertainment in their own right, they attracted fairgoers who otherwise might have attended some of the Agricultural Society's nightly meetings, at which the state's most progressive planters and agriculturists discussed the latest farming techniques. Members of the society considered such instructional meetings to be among the fair's most important features, and some expressed concern that the competing political speechmaking might lure away too many potential members of their audience. After all, papers on ditching, manuring, marling, liming, or "The Best Mode of Curing Bright Tobacco" could hardly compete with the stem-winding political harangues.

To entertain the large crowds and to improve box-office receipts, Raleigh's theaters offered their best attractions during fair week. Fair visitors in 1870 could spend their evenings after the fairgrounds closed at a performance of *Othello*. In 1873 two theaters presented shows, and in 1880 the famous touring

nightly meetings of the Agricultural Society. But although some fairgoers probably passed up the society's educational presentations for less intellectually stimulating activities, the opportunity for North Carolinians to see good theater often compensated for missed chances to learn more about scientific agriculture.

Enlivening Raleigh's social scene during fair week was a series of parties and balls that transformed the city into a continual gala. The Agricultural Society had established the custom, initiating an annual Marshals' Ball in the antebellum period. Early in the 1870s several Raleigh clubs and social organizations adopted the practice of holding parties during fair week. In 1874 Raleigh's Oak City Pleasure Club held a "grand hop" during fair week. The gathering, deemed a success, was repeated at the fair of 1875. Other Raleigh clubs soon followed suit. In 1876 the Monogram Club held a german, or cotillion, during fair week, and the Marshals' Ball and Pleasure Club's "hop" took place as usual. In 1881 the Agricultural Society added the Grand Coronation Ball to

Naturally, politicians took advantage of the large and influential audience the fair provided. Inasmuch as the orations of many political figures represented amusing entertainment in their own right, they attracted fairgoers who otherwise might have attended some of the Agricultural Society's nightly meetings, at which the state's most progressive planters and agriculturists discussed the latest farming techniques. Members of the society considered such instructional meetings to be among the fair's most important features, and some expressed concern that the competing political speechmaking might lure away too many potential members of their audience. After all, papers on ditching, manuring, marling, liming, or "The Best Mode of Curing Bright Tobacco" could hardly compete with the stem-winding political harangues.

company operated by John T. Ford of Washington and Baltimore played the city during fair week. Throughout the century, other well-known touring companies appeared in Raleigh during the week of the fair. The theaters, which were open at night, did not prevent fairgoers from participating fully in daytime fair activities. Like political speeches, however, they did lure people away from the

the already long list of social activities. Parties and balls continued to multiply throughout the 1880s, reaching a peak in 1892. In that year the city of Raleigh commemorated its centennial in conjunction with the fair. Among the parties and balls scheduled during the week were the Capital Club's german, the L'Allegro Club's german, the Grand Coronation Ball, and the Grand Centennial Ball. Those gatherings, representing the major events of a week of feverish social activity, received extensive press coverage. Although most were not directly connected with the fair, they considerably enhanced the growing social significance of fair week.

How fairgoers reacted to the crowds, bands, midway attractions, and galas of fair week is revealed by the accounts of participants. Peter Evans Smith, a Halifax County planter and engineer who helped construct several North Carolina railroads, spent part of his day at the 1870 fair and noted the effects of foul weather upon the crowd. "Over half the crowd," he wrote his wife, "got wet; and such a sight you never saw with umbrellas turned wrong side out—also a good many dresses in the same predicament." A reporter who attended the 1877 fair found that "Raleigh abounds in animation just now. Driving, sailing and promenading are in full play. The streets are an inspiring picture of life and activity. . . . The hotels are already thronged with humanity and there is no room for more. . . ." Another account, this one of the 1882 fair,

week's social events rather than to seek information about agricultural and industrial advancements. In the post-Civil War era, increased social activity in Raleigh during fair week likewise detracted from the fair's educational features. Social events in the city drew crowds away from the society's evening meetings at the fairgrounds—meetings at which the latest concepts in scientific agriculture were discussed. In addition to damaging the fair's formal instructional program, the whirl of contemporaneous social activity tended to undermine the educational value of the hundreds of exhibits displayed at the fair.

The state fair probably would not have survived had it not become a social attraction. Prior to the Civil War, it had been the state's only agricultural educational institution, other than the agricultural press, with a statewide impact. In the post-Civil War era, the growing influence of newly formed institutions such as the state Department of Agriculture, the North Carolina College of Agriculture and Mechanical Arts, and the state Agricultural Experiment Station, as well

> *Peter Evans Smith, a Halifax County planter and engineer who helped construct several North Carolina railroads, spent part of his day at the 1870 fair and noted the effects of foul weather upon the crowd. "Over half the crowd," he wrote his wife, "got wet; and such a sight you never saw with umbrellas turned wrong side out—also a good many dresses in the same predicament."*

captured the atmosphere of the city during fair week in these words: "The city has had a 'regular jamboree' for ten days past. The crowds and dust, ankle deep; the crowds and mud, knee deep; the crowded hotels, and the crowds that couldn't find a hotel; the crowds in search of water to drink; the crowds that drank liquor like water; the crowds of pretty girls; the crowds of handsome marshals; etc., etc., etc., etc.,—quantum stuff, enough of such stuff! . . ."

As the fair's social importance increased, however, the appeal of its formal agricultural instructional program declined. The more serious-minded agrarian reformers had recognized that fact as early as 1860. In that year the *North Carolina Planter* charged that too many Carolinians came to the fair to enjoy the

as improved agricultural periodicals such as the *Progressive Farmer*, were much more effective disseminators of information about the latest agricultural practices than was the fair. And even though such agencies recognized and even employed the fair as a means of reaching a larger audience, they tended to overshadow its educational significance.

During the nineteenth century, social distractions associated with the state fair increased dramatically, but the fair's social appeal never completely negated its instructional nature. Fairgoers who attended the fair primarily for the excitement of its social activity almost always saw exhibits of purebred livestock, agricultural implements, and improved seeds and fertilizers. Over time, those exhibits gradually supplanted the formal

During the nineteenth century, social distractions associated with the state fair increased dramatically, but the fair's social appeal never completely negated its instructional nature. Fairgoers who attended the fair primarily for the excitement of its social activity almost always saw exhibits of purebred livestock, agricultural implements, and improved seeds and fertilizers. Over time, those exhibits gradually supplanted the formal presentations offered at the nightly meetings of the Agricultural Society as the primary means of introducing scientific agricultural methods to Carolina farmers.

presentations offered at the nightly meetings of the Agricultural Society as the primary means of introducing scientific agricultural methods to Carolina farmers. The bands, the parades, and an increasingly commercialized midway served as enticements to attract more and more people to view the exhibits.

The North Carolina State Fair accurately reflected the social values of the state's citizenry in the second half of the nineteenth century. The popularity of military groups at antebellum fairs reveals a social mentality that could conceive not only of victory in the Civil War but also a quick and relatively easy one. Such events as the reviewing of troops by governors Zebulon Vance and Wade Hampton underscored fairgoers' devotion to the "Lost Cause" philosophy of the post–Civil War years, an era during which many who attended the fair had been participants in the Civil War. The pomp, color, and spectacle of the saber tournaments and grand coronation balls exemplified the devotion of nineteenth-century North Carolinians to a romantic vision—one that contributed both to their willingness to fight the Civil War and to continue to glorify the bravery with which they did so even after sustaining defeat.

Positive effects stemming from the social aspects of the nineteenth-century fair more than offset the damage done to the Agricultural Society's educational programs by social activities, however. The nineteenth-century fair provided a much-needed and appreciated week of entertainment and excitement for thousands of North Carolinians, especially those from rural areas. For many farm families, the fair was the major event in their social life, a welcomed interruption from the monotony and isolation of everyday existence. The state fair not only offered its own attractions; it also afforded rural and small-town North Carolinians a reason to visit the capital city. There they could see outstanding touring theatrical companies, listen to some of the state's leading political figures, and renew friendships and acquaintances. The fair also helped bind together people from different regions of a large and geographically diverse state in a period of time in which travel remained difficult, even with the development of major railroads and their relatively reliable timetables. By bringing together people from all parts of nineteenth-century North Carolina, especially those from the east and the west, and uniting them in a festive celebration of the state's agricultural and industrial advances, the state fair fulfilled one of the society's original expectations of it.

Social Activities at the Twentieth-Century Fair

During the first quarter of the twentieth century, previously established patterns of social activity both at the fairgrounds and in the city of Raleigh continued, inasmuch as the fair remained an exposition staged by a private organization whose membership included some of the state's most prominent and wealthiest citizens. Fair week had become even more than a significant aspect of the social life of the state, attracting the state's cultural elite to Raleigh. It also had become a celebration of who North Carolinians were, and an opportunity to express their dreams of what they hoped to be. Few events better illustrate this

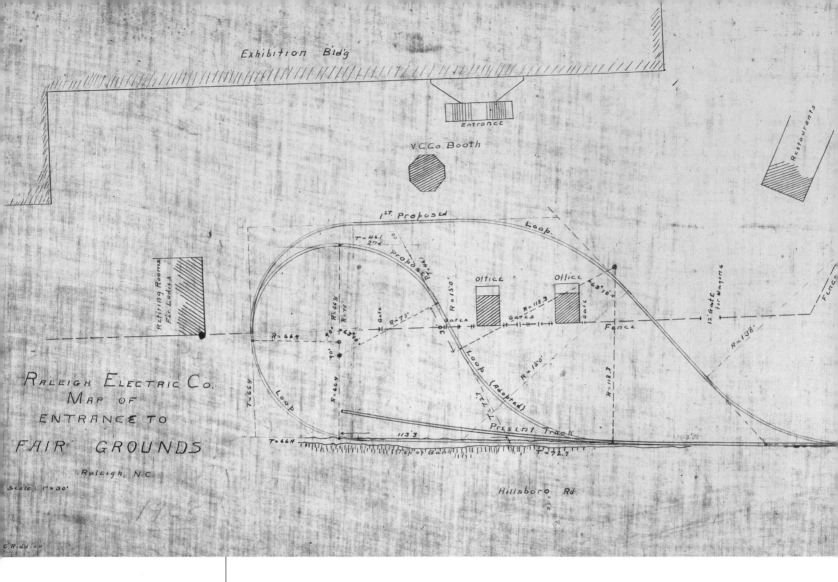

Map for a proposed trolley turnaround on the fairgrounds, which was constructed in 1908. This "loop" allowed increased trolley service between the fairgrounds and Raleigh, thereby making it more convenient for visitors in town for fair-week events. Image courtesy of the Wake County Register of Deeds.

reality than the creation of the North Carolina Literary and Historical Association in 1900. For the next century, that organization epitomized the state's aspirations to be recognized as a participant in, if not a leader of, the nation's cultural life. Affectionately referred to as "Lit and Hist" by thousands of North Carolinians who participated in its activities, the organization recognized the best literary and historical writing in the state in a given year and otherwise sought to encourage an appreciation of things cultural among the state's populace. One of its most significant contributions was its observation of "culture week" in Raleigh, a celebration of the cultural achievements of North Carolinians, usually held annually in November. Within a quarter-century after being established, the association achieved its ambitious goals of promoting the study of North Carolina history in public schools, building an important museum in Raleigh, and creating a superior library. In 1924 the society founded the *North Carolina Historical Review*, which remains one of the nation's most respected state historical jour-

nals. Now fallen upon hard times, in large part because of the success of various cultural organizations that it helped launch, the North Carolina Literary and Historical Association entered the twenty-first century with a declining membership, few funds, and little of the influence that it once had in the state's cultural activities. It seems a bit ironic that the North Carolina Literary and Historical Association, the arbiter of North Carolina's cultural standards, originated during state fair week, yet its origins reflect both the social position of the Agricultural Society's membership and the social appeal of the fair.

Most early-twentieth-century fair-week events were devoted not to high culture but to the more mundane concerns of a variety of organizations that scheduled their annual or semiannual meetings to coincide with the fair. Among such organizations was the Sons of Confederate Veterans, whose meetings reflected the loyalty of white North Carolinians not to the Confederacy per se but to a perception of the bravery of white Southerners on the battlefield and, crucially, to the concept of white supremacy. It must be remembered that the generation that led the Agricultural Society during the final quarter of the nineteenth century had actually participated in the Civil War. (For example, Julian S. Carr, a former president of the society active in its affairs as late as 1920, had been an officer in the Civil War; Carr greeted President Theodore Roosevelt dressed in his Confederate uniform when the chief executive paid a visit to the fair in 1905.)

Just as organizations within the state held their annual meetings at the fairgrounds, Raleigh organizations and citizens hosted various activities for the benefit of the crowds who converged on the city throughout the early twentieth century. During fair week, for example, Raleigh clubs continued to sponsor balls, to which the leaders of the Agricultural Society and other prominent visitors were invited. The Capital Club's german was a perennial favorite, although others, such as that of the Sphinx Club, captured press attention. The annual Marshals' Ball, held at Raleigh's City Auditorium for fair marshals and their wives, continued to be a favorite. In 1910, in recognition of the contribution of fair week to the city's economy, the citizens of Raleigh literally welcomed fairgoers into their homes. Prominent Raleigh residents provided guided tours of their homes, and hundreds of Raleigh citizens wore "Ask Me" badges to assure visitors that their questions would meet with a

friendly, informative response. Fairgoers also attended attractions scheduled especially for fair week at various venues, including the city's theaters. In 1910, for example, visitors of the fair could attend performances of a play titled "The Show Girl and the Academy" at a Raleigh theater. Visitors to the 1920 fair could choose from nightly shows at the Academy of Music, among them "Lasses' White Minstrels." Such shows, a reporter noted, were not strictly fair-related events, "but they come along during the week for the entertainment of State Fair visitors, and will pull additional thousands to the Agricultural and Industrial exposition."

College football also continued to compete for fairgoers' attention, with the football team of the North Carolina College of Agriculture and Mechanical Arts hosting an opponent on the Thursday of fair week. In 1900 A&M, as the college was often called, lost to Virginia A&M (now Virginia Polytechnic Institute) by a score of 18 to 2. During fair week in 1919, North Carolina State College of Agriculture and Engineering (as it was then known) met the University of North Carolina (UNC) for the first time since 1905. The State-Carolina fair-week game quickly became a tradition, but after 1925 UNC withdrew from the contest, leaving State to promise a match with a Southern Conference foe at future fairs. In the late 1920s, that foe turned out to be Wake Forest, but the State game scheduled

Traffic problems at the state fair increased markedly after the First World War, especially during weekends when North Carolina State College hosted football games. Shown here are cars entering the fairgrounds for the 1946 fair. Photo courtesy of A&H.

North Carolina State University's Carter Stadium (now known as Carter-Finley Stadium) was constructed adjacent to the fairgrounds in 1966. On Saturdays in October when State football games coincide with the state fair, traffic problems test the mettle of several police organizations. This aerial shot shows the resulting congestion at the 2002 fair. Photo courtesy of NCDA&CS.

on Saturdays during fair week soon became just another contest, especially with the relocation of the fairgrounds to more than a mile west of the State campus.

Despite the move, when games at State College's Riddick Stadium coincided with fair week, traffic could be a problem for the fair visitor. With the 1966 opening of North Carolina State's Carter Stadium adjacent to the fairgrounds, the complications of fair traffic and game traffic became at times nightmarish. That was especially the case if State just happened to be playing UNC, its arch-rival from down the road. Although thousands of North Carolinians continue to take advantage of the opportunity to attend a football game and the state fair in a single day's outing, many local residents and veteran fairgoers choose to avoid such traffic-plagued occasions. (Normally, however, North Carolina State Highway Patrol officers keep traffic moving smoothly when State's home football schedule and fair Saturdays coincide.)

With the state's acquisition of the state fair, the social scene in Raleigh underwent a profound change, the result primarily of two major contributing factors. First, as a state agency, the fair lacked the ambiance of a private organization whose leaders reflected the ideals and tastes of North Carolina's elite

during the height of the Victorian era. Instead, not surprisingly, the fair's more bureaucratic management began to encourage fair visitors to spend some of their time in Raleigh during fair week to visit the city's other state-operated attractions, as well as its educational institutions. In 1933, for example, the fair's management urged North Carolinians to attend the fair and to "Bring your boys and girls, let them also explore the Capital City, it will help them with their school work." "Take this occasion," North Carolinians were advised by fair management, "to see RALEIGH with its State Departments, Colleges, Schools, and various Other Centers of Interest," including the State Museum. For the fairgoer, balls, galas, and broad-saber tournaments were out, while visiting other state agencies and learning about their activities were in.

Dorton Arena and the Era of Professional Entertainment

By far the more significant factor in the demise of social activity in Raleigh directly associated with the state fair was the increasing diversity of entertainment offered at the fairgrounds. Especially with the construction of

Dorton Arena, which gave the fair a splendid concert venue, entertainment events scheduled for the city could no longer compete with the fair. With the new arena, the tables were turned: entertainment offered by the fair became an attraction for Raleigh residents—not the other way around.

With the advent of free performances by nationally known entertainers at Dorton Arena during the 1953 centennial fair, fair management inaugurated the practice of booking performers for what quickly became some of the fair's top attractions—free arena concerts and performances. While that practice has varied somewhat over the years, a basic formula continues to be followed: booking a slate of entertainers with a predominantly country music flavor, particularly those who emphasize "down home" or "traditional" values; including (after 1964) at least one black performer, who must be acceptable to a predominantly white audience; avoiding at all costs performances that could in any way be considered controversial. The very first performance in the new arena, which took place on Tuesday, October 20, 1953, set the pattern.

The show, billed as the WSM Grand Ole Opry Jubilee, included performers from Nashville's hallowed country music venue, individuals who would have been instantly recognizable to rural and small-town folks, as well as city dwellers originally from rural areas and small towns, who comprised the overwhelming majority of fairgoers. The show featured Hank Snow, one of the major country music stars of the time, and his Rainbow Ranch Boys; Cowboy Copas, another well-known country singer, with the Oklahoma Wranglers; and Kathy Copas; Randy Hughes; and Lazy Day Jim. It was, like the fair audience, an all-white show. It played the arena for four consecutive nights.

Other performances during the 1950s retained their country flavor but added performers introduced to the public by television, a new medium rapidly becoming the primary source of entertainment for the vast majority of North Carolina's families. For example, the 1956 Dorton Arena program included Pat Boone of the *Arthur Godfrey Show* and Russell Arms and Dorothy Collins from the enormously popular television show *Your Hit*

Since 1953, when the newly opened Dorton Arena gave the state fair an indoor venue for name performers, concerts have been one of the fair's most consistently popular attractions. Pat Boone, considered a "wholesome" representative of the newly emerging rock 'n' roll sound, played the 1956 fair.
Photo courtesy of *N&O*.

Roy Rogers, the "King of the Cowboys," and his wife Dale Evans, who played the arena in 1970, are typical of the all-American, down-home acts booked during fair week. Photo courtesy of *N&O.*

Parade, as well as the Midwestern Hayride, an Opry-like country music show. Noted television personalities who appeared at the fair during the 1960s included versatile musician Ray Charles, who performed pop tunes, country ballads, and blues numbers equally well and was one of the first black performers to appear in Dorton (1963), and Jack Bailey, star of the long-running radio-turned-television program *Queen for a Day*, who appeared in 1965. In 1967 the *Jimmy Dean Show*, fronted

by the popular country/pop performer Jimmy Dean, played the arena for a night and included performances by Boots Randolph, a country saxophone player, and the Cimarron Singers, a country group. That year a rodeo, which played the arena for four nights in a major deviation from the musical-performance format, was the featured attraction.

In the early 1970s, the Dorton Arena shows settled into the still-dominant routine—country music acts with a little variation for

the non-country audience. The 1971 programs provide a perfect example of the formula. That year Charlie Pride, a black country music star, was on the playbill; Paul Revere and the Raiders, a rock group, appealed to the younger crowd; Bob Hope, Margaret Whiting, and stars of the *Lawrence Welk Show* appealed to the older folks; and country music stars Roy Clark and Archie Campbell from the popular *Hee Haw* television show, as well as Ray Price and Jodi Miller, sang for Dorton's basic country music audience. The following year black pop stars Freda Payne and Al Green played at Dorton, as did soft-pop stars John Davidson, Tony Orlando and Dawn, and Bobby Vinton. Art Linkletter appeared as the featured television personality, and Mel Tillis, Ray Stevens, Donna Fargo, the Cornelius Brothers and Sister Rose, Tompall and the Glaser Brothers, and the Oak Ridge Boys were booked for country music fans. By 1974, fair management had added the final component to the Dorton Arena program format—a gospel group, in that year the Dixie Melody Boys, one of the most famous southern examples of the genre. From the mid-1970s on, the formula rarely varied. In 1979, for example, the arena country music acts featured Ronnie Millsap, a North Carolina native; Eddie Rabbit; the Kendalls, a father-and-daughter team;

Barbara Mandrell; and Don Williams. The Inspirations and the Dixie Melody Boys sang gospel, and a black group, Cornell Gunter's Coasters, sang their by-then nostalgic beach music from the 1950s.

Many artists played the fair with some regularity. Repeat performers in the 1980s included country comedian Jerry Clower, who returned in 1990 and 1993; country singers Ronnie Millsap and T. G. Shepard; Tammy Wynette, "the Queen of Country Music"; Ray Stevens, noted for his country novelty tunes; the Shirelles, an all-girl black rock group; the

Above: Fair management prefers to schedule homegrown talent at Dorton Arena. Here North Carolina-born country music star Ronnie Millsap performs at the 1996 fair. Photo courtesy of NCDA&CS.

Below: Since the initial concerts in Dorton Arena, country music acts have been by far the most frequently scheduled. Country music legend Willie Nelson performed at the 2000 fair. Photo courtesy of NCDA&CS.

Del Vikings, an old beach music group; the Village People, a pop/nostalgia group; the gospel group 4 Him; pop/country singer Billy Joe Royal; and country acts Diamond Rio, Rascal Flatts, the incomparable Loretta Lynn, and the Kentucky Headhunters played the arena. In recognition of the state's rapidly growing Hispanic population, the fair management added singer José Guadalupe Esparza to the playbill. In 2002 management instituted a five-dollar charge for Dorton Arena concerts, which had previously been free, in an effort to increase the fair's entertainment budget so that it could attract even bigger names to play the arena during future fairs.

Coasters, a rhythm-and-blues group popular since the 1950s; and gospel singers the Florida Boys and the Dixie Melody Boys. During the 1990s, black rock performers included the Shirelles and their always popular harmony; the Coasters; and Chubby Checker. The Florida Boys and the Dixie Melody Boys continued to be the most popular, or at least the most readily booked, gospel acts, along with country music performers Patty Loveless, Marty Stuart, Don Williams, Ronnie Millsap, the Oak Ridge Boys, and the hard-driving country rock Charlie Daniels Band, fronted by Daniels, a North Carolina native.

The arrival of a new millennium had no noticeable impact on the formula. In 2001 Lou Rawls, a gifted black pop/standard stylist; the

Horse Racing at the Twentieth-Century Fair

For much of the first half of the twentieth century, horse races drew fairgoers into the grandstands to see the spectacle of exciting racing, both harness and running events. In 1910, for example, the Raleigh *News and Observer* reported that "At four o'clock the splendid new grandstand was packed and the quarter mile—literally lined with people—a kaleidoscopic picture, with an interesting view at every turn of the eye." But horse racing at the fairgrounds remained as much a social event as a sporting contest, especially while the Agricultural Society sponsored the fair.

Racing clearly was part of Raleigh's social scene during fair week, with visiting horsemen receiving lavish attention. It was, for example, the custom for North Carolina's horsey set to treat "visiting horsemen" to "Delmonico Dinners" near the horse stalls between heats on the fairgrounds. (Such sumptuous spreads, attended by Raleigh's finest, obtained their name from New York's famed Delmonico's Restaurant, the haunt of the nation's financial and social elite.)

While horse racing attracted rural families to the fair and allowed North Carolina horsemen to race their prized animals against those of horsemen from throughout the East Coast, it remained an expensive proposition. Horse racing, both harness and running, required a well-maintained track and the stalls to accommodate the horses for up to a week. To promote a large, competitive, and attractive

field, the Agricultural Society had to struggle to maintain purses that would convince horsemen from outside the state to enter their animals. As in the antebellum era, the society's ability to deliver varied from year to year. In 1900 racing seemed strong, attracting a number of noted horsemen and their racing stock; by 1905, however, it was clear that racing was in trouble. In that year, in at least two harness classifications, the fair managed to attract but three entrants, hardly the number required to give spectators the thrills of competitive racing. The construction of a new grandstand and improved purses helped, and by 1910 racing once more drew a larger, more competitive field and remained robust into the 1920s. In 1921, the year following the election of Edith Vanderbilt of Biltmore, a noted horsewoman, as president of the Agricultural Society, the fair held its inaugural horse show. Staged to attract participants from North Carolina's agrarian elite, the horse show displayed the state's best riders on ponies, saddle horses, jumpers, and three- and five-gaited horses.

Soon after Vanderbilt's ascension to the presidency, the Agricultural Society decided to borrow funds and use them to construct a second racetrack and additional grandstands, both to improve the racing venue and to allow more room for the expanding midway. The facilities were completed in time for the 1922 fair. The 1925 fair opened with a daily program of both harness and running races for the entertainment of fairgoers. The annual horse show for that year, described as "The best display of fine horses ever," featured polo ponies and a demonstration of precision military horsemanship, both with officers from nearby Fort Bragg as riders. It soon became clear, however, that the society had made a crucial error in placing far too great an emphasis on horse racing and horse shows as a means of attracting a higher-toned social set. Neither the races nor the horse show provided the revenue required to meet the indebtedness incurred to construct the new racetrack and grandstands, which threw the Agricultural Society into a financial crisis from which it never recovered.

It soon became clear, however, that the society had made a crucial error in placing far too great an emphasis on horse racing and horse shows as a means of attracting a higher-toned social set. Neither the races nor the horse show provided the revenue required to meet the indebtedness incurred to construct the new racetrack and grandstands, which threw the Agricultural Society into a financial crisis from which it never recovered.

Two pacers drive for the finish line in front of the grandstands during a harness race at the 1938 fair. Photo courtesy of *N&O.*

When the state fair reopened in 1928 under state control, however, racing was back, as was the horse show. But it was racing that retained the crowds' attention. The 1928 fair featured a competitive field for both harness and running races. The harness races drew far more entrants than expected, among them some of the best horses in the East, including those from the stable of Dr. H. M. Parshall of Ohio, a dominant force on the harness-racing circuit. Harness racing struggled to survive during the 1930s as declining purses, the result of the depression, made it difficult to attract good horses. In 1929, for example, the fair had paid out $7,200 in purses; by 1932 that figure had declined to $2,500. Harness racing managed to retain its popularity during the 1930s, however, and Parshall continued his dominance of the racetrack during the decade. In 1930 he set a track record at the fair with a trotter, touring the track in two minutes and thirteen seconds. In 1935 he led a field of horsemen in trotting races (by that time under the sanction of the United Trotting Association, one of three sanctioning bodies in the United States). The 1935 harness field included entries from such well known North Carolina horsemen as W. N. Reynolds of Winston-Salem, as well as those from a number of other states. Running races, however, rapidly declined in popularity. In 1930 so few entrants appeared that they ran in unsanctioned races

under assumed names "for the enjoyment of the crowd," with no purses awarded. Unlike harness racing, running races at the fair fell victim to the depression and disappeared from fair programs by the mid-1930s.

With the formation of the United States Trotting Association in 1939, harness racing in the United States came under the control of one sanctioning body, and thereafter all such racing at the state fair came under its control. When the fair resumed in 1946 (following an interruption of four years caused by the Second World War), harness racing attempted to pick up where it had left off, but it could not compete with the forces of change in modern North Carolina. Interest in other organized sports, especially collegiate and professional team sports, rapidly outpaced the allegiance of an increasingly urban and suburban population to a sport that had its roots in the agrarian past. Gradually, fairgoers lost interest in even the three days of harness racing that were a standard feature of the fairs of the late 1940s and 1950s. In 1965 the state fair staged what it believed would be its last harness race. Noting that "Interest in harness racing has waned," the fair announced that "Harness Horse Racing, the oldest entertainment of the North Carolina State Fair, has been abandoned for a faster, more thrilling event. . . . The Jack Kochman Hell Drivers, starting with Monday afternoon, will produce their thrill

Granny's Oven, racing out of Ohio, captured the championship at the state fair's final harness racing series, staged at the 1996 fair. Commissioner of Agriculture Jim Graham presents the winner's trophy. Photo courtesy of NCDA&CS.

Although horse racing declined in popularity in the post–World War II era, the State Fair Rodeo become a major attraction by the 1970s. Pictured here is barrel racing at the 1984 rodeo event. Photo courtesy of NCDA&CS.

show event twice each day, and replace the Tuesday, Thursday, and Friday harness horse events." Eventually, nostalgia for the old harness races prevailed, and in 1991 the fair's management revived them. Nostalgia could not fill grandstand seats, however, and in 1996 the fair held its last harness race, signaling an end not only to an era but also to the attraction's nostalgic appeal.

Folk Festivals and Nostalgia

Nostalgia became one of the special ingredients of fair week, especially after the Second World War. As the state underwent rapid industrialization and urbanization during the last half of the twentieth century, the fair staged exhibits and programs that recalled its citizens' agrarian heritage. In 1948, under the direction of Bascom Lamar Lunsford, the fair held its first folk festival before a grandstand audience. Lunsford, a North Carolina legend, was born into a musical family in Madison County. At the age of thirteen, Lunsford, instructed by his father, a teacher at Mars Hill College, and an uncle, was already an accomplished fiddler and banjo picker. After being schooled in several private academies and at Rutherford College, he found it difficult to

adapt to any one occupation, and he worked at various times as a schoolteacher, a nurseryman, a lawyer, a college instructor, an auctioneer, and a newspaperman—achieving neither success nor distinction in any field. He married Nellie Triplett in 1906, and in 1925 Lunsford and his family settled on his wife's Buncombe County farm. There Lunsford devoted himself to what he believed to be his true calling: the preservation of old-time mountain music. Influenced by famous folklorist Robert Gordon, Lunsford collected some three thousand songs and related material from his native region and in 1924 recorded many of them. Unlike some of his contemporaries who became recording stars, Lunsford never became a commercial success, despite penning in 1920 the song "Mountain Dew," which subsequently became not only a popular country tune in praise of the virtues of homemade corn liquor but also a virtual musical synonym for North Carolina's hill people, known for their affinity for that product.

In 1928 Lunsford arranged his first folk gathering, the Mountain Dance and Folk Festival, in Asheville. He devoted the remainder of his life to arranging, promoting, and conducting such festivals. Driving battered old automobiles, he toured North Carolina's Appalachian region in search of performers for his festivals. Lunsford insisted that the festivals present authentic folk songs and dances: he allowed no commercial costumes or "flat foot" dancing and favored the use of acoustical (unamplified) music (although some amplification was accepted). He worked with New Deal agencies during the depression to bring his beloved music to local folk festivals, and in 1939 Franklin D. Roosevelt invited him to the White House to perform for King George VI and Queen Elizabeth of Great Britain.

J. S. Dorton encountered Lunsford at folk festivals in Asheville and Chapel Hill and felt the program would play well at the state fair. Dorton, as usual, was correct. When he announced the fair's first folk festival in 1948, he

From its beginning in 1948, the State Fair Folk Festival, a celebration of traditional song, dance, and storytelling, was an instant success. These traditional musicians performed with non-amplified instruments at the folk festival around 1950. Photo courtesy of NCDA&CS.

declared that it had two purposes. The first was "to provide additional high type and wholesome entertainment for fair visitors, especially to those of our own state and their children who have a material and cultural interest in the Fair Program." The second was to "encourage and foster efforts among the people of the state to preserve and develop our own rich heritage of native folk arts in the forms of balladry, folk song, Gospel singing, traditional folk games and dances, and string music; to the end that the benefits derived from seeing this material presented at its best may be extended to other interested recreational and social groups throughout the state who prize its value." The 1948 State Fair Folk Festival was an instant hit with fairgoers and became a feature of every succeeding fair program.

Although it has sometimes changed venues, the festival's format remained and remains essentially the same—native North Carolinians and natives of the Mountain regions of the South singing, playing, and dancing to the authentic folk music of their homeland. The categories in which contestants were entered at the 1980 folk festival, which played Dorton Arena three times a day except Sunday, are typical. They included clogging, smooth dancing, country and western music, bluegrass bands, gospel singing groups, fiddlers, banjoists, and balladry. The folk festival continues to be one of the most

Above: Traditional musicians performing at the folk festival during the 1984 fair (albeit with amplified instruments). Photo courtesy of NCDA&CS.

Right: Folk festival music remains traditional and basic. A cartoon cow's head (far right) watches the house band perform under the folk festival tent at the 1996 fair. Photo courtesy of NCDA&CS.

popular events at the state fair, a reminder and a celebration of a heritage of which residents of the entire state, not just the Mountain region, are justly proud.

The immediate popularity of the folk festival prompted J. S. Dorton to launch another nostalgic exhibit in 1951. State fair management inaugurated the "Village of Yesteryear" as an educational division "with the hope of renewing interest in our age-old arts and

crafts." The department, according to fair management, "features craftsmen at work in the many varied crafts from wood carvings to weavings, and basketry to candle wicking." Entry was open to "all North Carolinians and Southern Highland area craftsmen." Like the folk festival, the "Village of Yesteryear" was an instant success. The 1953 centennial edition of the fair featured it as an exhibit "where fair visitors will have an opportunity to see how

farmers have worked and lived and progressed over the span of a hundred years." Behind a split-rail fence, fairgoers could see "men working on chairs, rugs, pottery, and other handicrafts now almost extinct except in remote communities." By the mid-1960s, the Village exhibit included a number of old farming tools and machinery, among them "old foot-tredle churns, pewter pots, and horse-drawn reapers that progress have [*sic*] outmoded." In 1966 fair manager Art Pitzer

employed a blacksmith-carpenter to restore the tools and machines in the collection to working condition. Catching the wave of the post–World War II nostalgia boom, the Village expanded to exhibit the works of craftsmen and women and quickly required more space. In 1974 fair management described it as "one of the Fair's most popular and crowded attractions." That year the fair opened an Old Farm Machinery Building to provide exhibit space for its growing collection of antique farm

A weaver demonstrates the use of the handloom in the "Village of Yesteryear" at the 1984 fair. Photo courtesy of NCDA&CS.

A blacksmith plies his trade in Heritage Circle at the 1984 fair. Photo courtesy of NCDA&CS.

A demonstration of log splitting at the 1983 fair. Photo courtesy of NCDA&CS.

The schoolhouse in Heritage Circle, where each year Paul Blankinship mounts an exhibit on the history of the state fair. Photo (1983) courtesy of NCDA&CS.

tools and machinery. In 1975 the continuing popularity of crafts exhibits gained them an exhibit space of their own—the Gov. Holshouser Building, which remains the center for all craft exhibits at the fair. Inside the modern circular structure, fairgoers admire, and sometimes purchase, items made by potters, tinsmiths, toy makers, candlemakers, weavers, basket makers, and a host of others. It remains one of the busiest exhibit halls on the modern fairgrounds.

In 1974 fair management introduced the concept of a Heritage Circle, a "historically-oriented theme area adjacent to the fairgrounds lake" on the west side of the campus. The Gov. Holshouser Crafts Pavilion (named for Gov. James E. Holshouser Jr.), which the state legislature had appropriated $310,000 to construct, was the central feature, but the Circle was also designed to accommodate a variety of authentic structures representative of the state's rural and agrarian past. Items from three major areas of cultural interest overseen by the North Carolina Department of Cultural Resources (the Arts, Archives and History, and the State Library), displayed in a "relocatable dome called the Charter-Sphere," composed the initial exhibit, the focal point of which was the "Buggymobile," built by Gilbert Waters of New Bern in 1903 and the first automobile manufactured in North Carolina. (The exhibit also includes additional "artifacts depicting state heritage.") The first historical structure to be placed in the Heritage Circle area was an authentic nineteenth-century schoolhouse donated in 1975 and presently used to house an exhibit on the history of the state fair mounted each year by Paul Blankinship, a longtime collector of photographs, images, and memorabilia. Since its beginning, Heritage Circle has grown to include a church (which is used every day of the fair for worship services and gospel sings), a log cabin, and a tobacco barn, all authentic structures moved to the fairgrounds from their original locations. It also contains two structures built on the site from original materials: a molasses shed/cabin and a craft building. New structures built on site to resemble old buildings include a cabin equipped with a cider press and a blacksmith's workshop.

Not all efforts at creating ongoing attractions based on nostalgia succeeded, however. In 1955 fair management inaugurated the "singing conventions." In North Carolina and throughout the South, such gatherings, which featured gospel groups singing in close harmony, had taken place since the pre-Civil War years. The conventions hosted at the fair lasted only through 1961. Perhaps they failed to attract the crowds that flocked to the "Village of Yesteryear" because they were not nostalgic enough—that is, because gospel singing, traditionally popular in the state, was seen as something North Carolinians actually did and not something they merely watched others do.

The death of fair-hosted singing conventions by no means reflected a general demise in fairgoers' demand for musical entertainment, so much a part of the nineteenth-century fair experience. Bands remained a popular attraction of the twentieth-century fair, although the nature of the musical organizations changed considerably. In the first quarter of the century, when the Agricultural Society ran the fair, the band of the

A gospel group performing hymns in front of the church in Heritage Circle at the 1999 fair. Photo courtesy of NCDA&CS.

Third Regiment of the North Carolina National Guard, based in Raleigh, remained a favorite, as did the band of the State School for the Blind. Later in the period, the "Regular Army Band" of the Twenty-eighth Infantry, North Carolina's allotted regiment in the regular United States Army, played for fair crowds. The North Carolina State College Marching Band entertained crowds at the 1928 fair. During the 1930s, George Hamid frequently brought a band with his shows, usually "Cervone's Band." With the revival of the fair after the Second World War, bands, particularly the military variety, continued to be a popular feature.

Politics at the Fair

Politics has always played a role in the fair's success, and presidential politics during the twentieth century became an even more significant element of fair week. In 1896, with Democrat William Jennings Bryan and Republican William McKinley locked in a ferocious battle of competing ideologies, leaders of the Agricultural Society realized that an appearance by Bryan in Democratic North Carolina would attract thousands to the fair. The society attempted, without success, to persuade the nominee to attend. Nine years later, however, the first incumbent president to visit the fair appeared in the person of Theodore Roosevelt. A hugely popular Republican, Roosevelt visited during a year in which there was no presidential election in an effort to gain support for proposed federal regulation of railroads, a proposition that enjoyed widespread support in North Carolina and the rest of the South. Roosevelt arrived at the fairgrounds from downtown Raleigh at the head of a large parade that included many African Americans, among them students from Raleigh's Shaw University—a suggestion of the black population's continuing allegiance to the party that had freed it from slavery only forty years earlier. Roosevelt addressed a crowd of more than twenty-five

Theodore Roosevelt was the first president to speak at the North Carolina State Fair (1905). Although a Republican, the progressive, militantly expansionist Roosevelt was popular throughout the South. Photo courtesy of A&H.

thousand people, pleading for support of railroad regulation and the creation of a proposed Appalachian Forest Preserve (which eventually became the Great Smoky Mountains National Park). Roosevelt failed, however, to mention the issue of race to an audience that included Democratic politicians who had only recently disenfranchised and segregated the state's African American population.

Presidential visits to the North Carolina State Fair resumed after the Second World War amid hotly contested races in which Democrats sought to ensure that North Carolina remained a part of the Solid South in casting its electoral votes for the party's candidate. Such tight races usually produced a major national figure who touted the virtues of the Democratic presidential candidate. In 1948 incumbent president Harry S. Truman battled not only his Republican opponent Thomas Dewey but also Henry Wallace, running as the Progressive candidate and, a far more serious threat in North Carolina, Strom Thurmond, governor of South Carolina, running as a Dixiecrat. The Dixiecrats, breakaway southern Democrats angered at Truman's support of

civil rights legislation and his desegregation of the armed forces and federal civil service, threatened to carry much of the South, including North Carolina. Truman elected to come to the state fair to make his case to the people of North Carolina. Truman arrived at the fair by motorcade, accompanied by North Carolina's most influential Democrats, including Gov. R. Gregg Cherry and W. Kerr Scott, who had recently resigned as commissioner of agriculture to make a successful bid for a U.S. Senate seat. Accompanied by his wife Bess and daughter Margaret, Truman attacked Republican economic policies, especially the party's farm policy, to the delight of the more than seventy-five thousand people in attendance. E. James Strates remembers bringing Truman a teddy bear "for his grandchild." Truman, Strates recalled, looked him straight in the eye and said, "I have three grandchildren." Strates promptly produced two additional teddy bears. Truman subsequently carried North Carolina but lost the states of the Deep South.

During the John F. Kennedy-Richard M. Nixon contest of 1960, when it appeared that Nixon might carry the state, Kennedy sent the

An embattled Harry S. Truman came to the North Carolina State Fair in 1948 to persuade North Carolinians to reelect him over Republican opponent Thomas E. Dewey. Truman's appearance began a series of visits to the fair by presidential aspirants or their representatives during close election contests. Photo courtesy of A&H.

enormously popular former president Harry Truman to speak at the state fair on his behalf. Truman spoke to a packed audience in Dorton Arena on October 13, a day after Frederick Mueller, U.S. secretary of commerce, had addressed the crowd in support of Nixon's candidacy. Kennedy barely carried the state, winning one of the closest presidential contests ever. In 1976, in company with North Carolina's first Republican governor of the twentieth century, James E. Holshouser Jr., Gerald R. Ford visited the fair. Ford's appearance reflected the growing influence of the Republican Party in North Carolina and the

Above: Truman returned to the fair in 1960 to speak on behalf of Democratic presidential hopeful John F. Kennedy, who was locked in a close race with Vice-President Richard M. Nixon. Seated in the front row behind Truman, left to right, are Gov. Luther Hodges, North Carolina commissioner of agriculture L. Y. Ballentine, and Democratic gubernatorial candidate Terry Sanford. Photo courtesy of *N&O.*

Right: By the 1970s, growing economic prosperity, shifting demographic patterns, and the political fallout of the Civil Rights movement had firmly established the Republican Party as a political power in North Carolina and throughout the South. As a result, Republican presidents began to appear at the state fair during hotly contested elections. Gerald R. Ford sought the support of fairgoers at the 1976 fair during his contest with Democrat Jimmy Carter. Photo courtesy of *N&O.*

rest of the South. It also reflected the importance of North Carolina's relatively large electoral vote, for Ford was in a heated contest with Gov. Jimmy Carter of Georgia, the Democratic nominee. Ford hoped to repeat Richard Nixon's 1972 victory in the state, employing what political analysts called the "Southern Strategy." Speaking in Dorton Arena, Ford promised to cut taxes and balance the budget. James T. Broyhill, a Republican congressman from North Carolina, shared the platform with Ford, Governor Holshouser, and the Democratic commissioner of agriculture, Jim Graham, who appeared "out of courtesy" and "buried his head in his hands or looked at the floor" as the president spoke. Ford proved unable to counter the appeal of a native southerner and lost the state by a whisker to Carter.

In 1992, yet another incumbent Republican presidential candidate appeared at the fair in tandem with the state's second Republican governor. Like Ford, he was a northerner in a heated race with a Democratic candidate who was a governor of a southern state. George Bush, flanked on the platform by the state's Republican establishment, including Gov. James G. Martin and U.S. senator Jesse Helms, addressed an outdoor crowd of some seventeen thousand. Bush desperately sought to counter the impact of an economic slowdown that was quickly eroding the popularity he had enjoyed only a year earlier, reminding the crowd that while the United States was in a global recession, its economy was faring better than most others throughout the world. Bush ended his remarks with an appeal for the re-election of Martin and the election to the United States Senate of Republican candidate Lauch Faircloth. Smarting from the recession and unimpressed by the relative strength of the American economy when placed in global

Republican president George Bush appeared at the 1992 fair during his unsuccessful campaign against Democratic challenger Bill Clinton. Photo courtesy of the George Bush Presidential Library.

perspective, North Carolinians again placed their faith in a native southern Democrat, Bill Clinton, Democratic governor of Arkansas, who eked out a victory in the state and nation. As president, Clinton, too, visited the state fairgrounds, but not during fair week. On September 14, 1996, Clinton came to North Carolina to view the ravages of Hurricane Fran, which had devastated large areas of the state's coastal and central regions. During his visit he held a press conference at the fairgrounds. He returned there early in November 1998 to speak to a gathering of North Carolina's Democratic faithful in support of United States Senate candidate John Edwards.

Prominent state politicians, of course, had long campaigned for votes at North Carolina State Fairs of the twentieth century. Until 1925, the final year the North Carolina State Agricultural Society staged the fair, the annual fall event opened with a parade of dignitaries from downtown Raleigh to the fairgrounds. In that year, Gov. Angus McLean, society president and future governor O. Max Gardner, and their wives rode to the fairgrounds in a "bright and aristocratic Lincoln," followed by Raleigh "Boosters," the Shrine Club drum corps, several hundred State College cadets, and, dominating the procession, "a brilliant

array of floats designed by college students." Indeed, according to the *News and Observer*, all other parade units were merely "automobiles full of masculine faces with only an occasional woman to add anything of beauty." The incumbent governor, unless absent because of unavoidable scheduling conflicts, normally spoke to fairgoers at the opening ceremony, always assuring them that they lived in a great, progressive commonwealth. When the state acquired the fair in 1928, the parade was abandoned, but the governor continued to preside at the fair's formal opening. In that year Gov. Angus McLean, "accompanied by the red coated State College band and by staffers and directors of the fair," addressed that portion of the twenty thousand people on hand who could crowd into the grandstands. McLean, in a typical political speech, provided "a brief history of the State Fair as it was formerly held under an independent body," then assured the crowd "that he had always favored a state owned and operated institution." At times a governor might engage in a folksy stunt to establish his knowledge of the state's agricultural economy, as in 1951, when Commissioner of Agriculture L. Y. Ballentine and Gov. W. Kerr Scott engaged in a milking contest. Ballentine won, coaxing

President Bill Clinton appeared at the fairgrounds when he visited the Raleigh area, but not during fair week. Clinton, second from right, discusses the ravages of Hurricane Fran in 1996.
Photo courtesy of NCDA&CS.

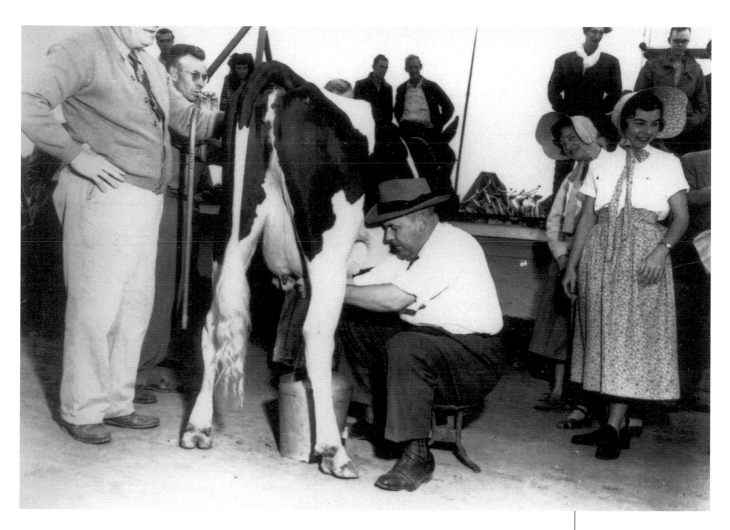

5.6 pounds of milk from his cow in two minutes, to only 3.3 pounds for the embarrassed dairyman Scott.

With the construction of Dorton Arena, fair officials moved the formal openings indoors. Opening ceremonies continue to be held there unless some special occasion dictated otherwise. In 1983, for example, Gov. James B. Hunt Jr. opened the brand-new horse complex that bears his name. Rarely were opening-day crowds privileged to hear great oratory on significant topics. Governors, understandably, stuck to traditional, upbeat speeches that praised the common man, as inclusively defined as possible. Typical were the following remarks delivered by Gov. Robert W. Scott at a special photograph exhibit sponsored by the North Carolina Arts Council on the day before the formal opening of the 1970 state fair: "The Arts Council serves to remind us that the arts have down-home relevancy, that the arts are not exclusively for the 'elite,' or for high society, that there is a universal appeal and meaning to the arts for everyone, rich and poor, rural, urban and suburban, for black, white, and Indian." A modern governor, well aware of recent changes in

the makeup of the state's population, might expand Scott's list to include at least Hispanics and Asians.

With the steady growth of the Republican Party in North Carolina after the Second World War, both Democrats and Republicans

Above: Commissioner of Agriculture L. Y. Ballentine milks a cow at the 1951 state fair. Photo courtesy of A&H.

Left: Political booths at the state fair reflected the social and political turmoil of the 1960s, especially the passions generated by the Civil Rights movement and the Vietnam War. Pictured here is a booth sponsored by the Ku Klux Klan at the 1966 fair. Photo courtesy of N&O.

Right: A booth sponsored by the John Birch Society at the 1966 fair urged the impeachment of Earl Warren, chief justice of the Supreme Court of the United States. Photo courtesy of *N&O*.

Below: The Republican Party sponsored this booth at the 2002 fair. Note the poster for Elizabeth Dole, who visited the fair in her effort to succeed Jesse Helms as United States senator. Photo courtesy of the author.

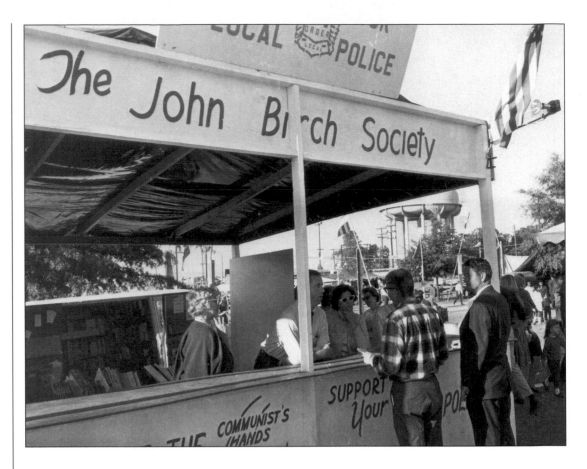

used the fair to persuade the thousands who attended each year that their party's particular political vision best assured the well-being and prosperity of the state's citizenry. To do so, both parties established exhibit booths from which they distributed bumper stickers, lapel buttons, and campaign literature in support of their candidates. By the 1960s that political competition, sometimes quite heated, continued even in odd-numbered years, when there were no presidential or statewide races.

Each party sought some means of attracting the attention of potential voters. In 1965, for example, the party faithful at the Republican booth urged voters to complete a "ballot" on four items, all designed to capture the attention of fairgoers. The ballot contained questions that were a mixture of the serious and the humorous, including one on whether "Beatle" haircuts should be allowed. In addition to the major political parties, other organizations with political agendas likewise began to employ exhibits at the fair as part of their strategy to influence North Carolina voters, especially in the politically and socially turbulent sixties. For example, during the integration battles of the early 1960s, the Ku Klux Klan expounded its white supremacist views from a booth at the fair. In 1965, in less emotionally charged circumstances, workers at the John Birch Society booth distributed literature urging the United States to get out of the United Nations and that Earl Warren, chief justice of the United States Supreme Court, be impeached. Currently other political parties, such as the Libertarians, compete with Democrats and Republicans for the attention and votes of fairgoers.

Politicians regularly "worked the crowds" during fair week, shaking hands, hugging babies, and handing out literature. Jim Graham was the past master at using the state fair to advance his political career. During fair week he was always visible patrolling the grounds either on foot or, as he aged, in a chauffeured golf cart, promoting himself, North Carolina agriculture, and the fair itself. Graham, of course, had a huge advantage because his department ran the fair, but other politicians followed his lead. The fair provided the perfect venue for meeting large numbers of people in a short period of time, and on occasion working the fair may have decided important political contests.

The politics of the state fair are every bit as intriguing as politics at the fair. During the nineteenth century, the leaders of the Agricultural Society, because of their wealth and prominence, had enormous political clout. Some, such as Thomas M. Holt, who served as governor from 1891 to 1893, and Thomas Ruffin, a member of the state supreme court, were among the state's political elite. Little changed in the twentieth century. In fact, much of the politics of the twentieth-century fair is encapsulated in the history of one remarkable political family—the Scotts of Alamance County.

The relationship between the Scotts and the state fair began in 1901, long before the fair became a state agency, when Gov. Charles B. Aycock appointed Robert Walter Scott, a noted dairy farmer from Alamance County, to the State Board of Agriculture. Scott, a graduate of the University of North Carolina, served in the North Carolina House of Representatives from 1889 to 1893 and as a state senator in 1901 and 1929. He served as a member of the Board of Agriculture until his death in 1929, a position that allowed him to encourage the increasing involvement of the Department of Agriculture in the operations of the state fair during the first quarter of the twentieth century. Furthermore, Scott endorsed the state's acquisition of the fair in 1928. Robert Scott's son, W. Kerr Scott, in his successful 1936 bid for the office of commissioner of agriculture, employed the issue of making the state fair a

division of the Department of Agriculture in 1936. As commissioner from 1937 to 1949—and later as governor from 1949 to 1953—W. Kerr Scott made the inspired choice of J. S. Dorton to manage the fair and also consistently supported Dorton's ideals for an expanded fair. Scott's protégé, Assistant Commissioner of Agriculture L. Y. Ballentine, became commissioner in 1949 and remained in that position until his death in 1964.

In 1964 Jim Graham replaced Ballentine as commissioner. Graham, too, was a W. Kerr Scott protégé, and Graham, like Scott, was a graduate of North Carolina State College and a cattleman. Commissioner Scott assisted Graham in obtaining his first job with the Department of Agriculture—head of the Ashe County Test Farm. Gov. Terry Sanford, in many ways W. Kerr Scott's political heir, rewarded Graham for his support in the 1960 gubernatorial campaign by appointing him commissioner. Graham returned the favor in 1964 by supporting the unsuccessful progressive Democratic candidate, Richardson Preyer, in a difficult three-way gubernatorial Democratic primary fight. As commissioner, Graham had the good fortune to serve under

Robert W. Scott as governor in 1969. He is the father of the former commissioner, Meg Scott Phipps. Photo courtesy of A&H.

two governors who graduated from North Carolina State, the first being Robert W. Scott, W. Kerr's son, who served from 1969 to 1973, and the second being Jim Hunt. Just as his father had helped Dorton, "Bob" Scott supported Graham's request for state funds to allow the fair to develop along plans Dorton had formulated in 1946. Graham, meanwhile, consistently aligned himself with the progressive wing of the Democratic Party—those Democrats who felt that the state should continue to expand the services that it provided to its citizens, especially in the realms of education and economic development.

Graham, like the Scotts, firmly believed that the Department of Agriculture should support and speak for North Carolina's agricultural interests. He also felt strongly that the department should never fall under Republican control (he even claimed to have decided not to run for governor as a means of ensuring his election as commissioner). It seemed appropriate that the person who replaced the retiring Graham as commissioner of agriculture in 2001 was Meg Scott Phipps, the daughter of Bob Scott, granddaughter of

Jim Graham with Gov. Daniel K. Moore at the state fair in 1967. Photo courtesy of *N&O.*

W. Kerr Scott, and great-granddaughter of Robert Scott. Circumstances surrounding campaign finances and departmental contract awards, including the carnival operator for the 2002 fair, prompted Commissioner Phipps to resign her office at mid-term.

Although a fair manager, assisted by a professional staff, has headed the state fair since 1937, North Carolina's commissioner of agriculture remains the single most influential officeholder in terms of seeking support for the fair from the governor and the state legislature. In an increasingly partisan climate, it is quite unlikely that any future commissioner will enjoy the tenure of office attained by Jim Graham. That climate also increases the possibility that future commissioners could face a

state legislature or governor, or both, controlled by the opposition party, which might make it more difficult to secure favorable responses to measures advanced by the fair's management.

Uniting North Carolinians

Throughout its entire existence, the North Carolina State Fair has served to unite the people of a large and geographically diverse state through a common experience. That goal, forcefully stated in the 1853 charter of the North Carolina State Agricultural Society, was equally important to the civil servants who ran the fair after 1928. In his address to the

North Carolinians have always journeyed to the state fair to see other North Carolinians from different parts of the state, as the fair's founders intended. Being in the crowd is a large part of the fair's appeal. Shown here are throngs at the 1946 fair. Photo courtesy of *N&O*.

opening-day crowd of the 1928 fair, Gov. Angus McLean observed: "Our state has three major sections and a wide variety of products. It is, therefore, most important that our own people should become better acquainted with these things, so that we may fully understand our own resources and the potentialities they afford." McLean also "stressed the importance of cooperation among the people in promoting the State Fair and the things the fair stands for . . . especially in view of the fact that the fair has now been made a state agency and has no local aspect whatsoever. . . ." In 1937 recently appointed fair manager J. S. Dorton wrote that it was the fair's purpose to serve every section of the state and "to reflect the people of the State and the accomplishment of all fields of endeavor." Speaking at the 1946 fair (the first to be held since before World War II), Gov. R. Gregg Cherry noted that "we can once again present to the people of North Carolina that old time-honored institution,

the North Carolina State Fair, which in the past contributed so much to the entertainment and general well-being of Tar Heels everywhere." Commissioner of Agriculture W. Kerr Scott echoed those sentiments at the official reopening of the fair in 1946: "At the State Fair we can meet together and talk together. . . ." Writing of the 1953 centennial fair, Commissioner of Agriculture L. Y. Ballentine observed that the state fair "brings together, for a better understanding of each other's problems, the rural and urban population of the state."

At the 1961 fair, management conducted a survey to determine where fair visitors resided. The results revealed that to a remarkable degree the fair had become an institution for North Carolinians from all sections of the state. While Wake and its adjacent counties furnished 40 percent of that year's fairgoers, the majority of the visitors came from an additional sixty-two counties, and a few came from out of state. Thirty-six percent of those surveyed identified themselves as rural residents, 22 percent were "city folks" from North Carolina's seventeen towns with a population greater than 23,000, and 24 percent hailed from the state's numerous small towns. In 1996 a similar survey revealed that the fair continued to draw visitors from throughout the state. While Wake and adjacent counties contributed 47.3 percent of all those attending, every other county in the state was represented in the attendance figures; and 8.3 percent of those in attendance came from Virginia, South Carolina, or Georgia. The poll,

Above: A part of the fair that most fairgoers never see includes the camps of the midway workers, who set up a miniature city at each state fair they serve. Pictured here are midway worker quarters at the 2002 state fair. Photo courtesy of the author.

Left: Crowds at the 1996 fair. Photo courtesy of NCDA&CS.

conducted only among adults, also revealed that the fair attracted people of all ages, with 43 percent of those in attendance in the 18-to-34-year-old age bracket, another 44 percent in the 35-to-54 segment, and 15 percent in the over-55 category. The fair had become, as its founders had dreamed more than a hundred years before, a sprawling, colorful, multi-faceted, and indispensable institution in the social life of North Carolina.

Institutions, of course, are created and sustained by the people whose needs they fulfill. The North Carolina State Fair became such a significant institution in the life of North Carolinians because it met their social as well as their educational needs. For a few days in October, the fair allowed North Carolinians to boast of their agricultural and industrial progress, to learn of the latest in agricultural techniques, to see how people in other sections of the state lived, and to enjoy themselves thoroughly while doing so. The North Carolina State Agricultural Society's concept of an instructional fair, modified by the public's demand for a festival, resulted in the creation of a genuinely unique institution that became and promises to remain an integral part of the state's social life for many years to come.

Reaching and Teaching Farm Families

Agricultural Promotion at the Fair

Promotion of Agriculture at the Nineteenth-Century Fair

The North Carolina State Agricultural Society established the state fair primarily to improve agriculture in North Carolina, and the promotion of scientific agriculture remained the fair's major objective throughout the nineteenth century. The fair concentrated on the promotion of no particular phase of agricultural production but instead attempted to improve every aspect of the state's agricultural life. It encouraged improvements of livestock, field crops, vegetables, farm implements, and farming methods. The society employed essays and speeches in its efforts to educate North Carolinians to the need for improved agricultural practices, but with little success. With its premium lists, the fair encouraged displays of the best of the state's agricultural products and practices, which helped to educate North Carolina farmers to the benefits scientific agriculture held for them. Displays of the latest agricultural implements and products such as fertilizers, exhibited both by manufacturers and retailers, likewise informed farmers of how they could improve their agricultural practices.

Canned goods with ribbons at the 1996 fair. Photo courtesy of NCDA&CS.

and Devons. Purebred cattle, in the form of a registered Shorthorn bull, were introduced in 1841. Livestock exhibits at antebellum fairs reflected North Carolina's lack of purebred cattle. The antebellum fair required no pedigree of any animal exhibited in any class, for such a requirement would have eliminated cattle exhibits. Furthermore, the fair offered premiums not for each breed but for *types* of cattle, such as milch cows, mixed blood and native cattle, and working oxen, at times placing several breeds, such as Devons, Ayshires, and Holsteins, in one class. The fact that such breeds were named in the fair's premium lists suggests that some farmers displayed cattle with enough breed-specific traits to be classified as such by the judges, but none of the entries were registered animals.

Fair officials continually sought to improve livestock exhibits in an effort to introduce North Carolina farmers to blooded animals. It offered a premium for cattle from other states and required pedigrees of animals entered in that class. To further encourage farmers to exhibit their cattle, the fair paid the feeding costs of all cattle exhibited. In 1873 the fair began offering premiums for breeds rather than types of cattle; examples of such breeds included Devons, Durhams, Ayshires, Alderneys, and Brahmas. Still, no pedigree was required of exhibits in any class; rather, judges determined the class in which each animal would be shown. Judges based their decision to enter a specific animal in a breed-premium classification by viewing the animal in question and determining which breed it most resembled. Under that rather primitive practice, one judge, obviously not well acquainted with blooded animals, actually refused to enter a blooded Devon in an exhibit because he did not think the animal looked like a Devon. In 1873 the fair also obtained exhibits of registered, purebred cattle from outside the state, including a Dutch Belted bull from Delaware, Jerseys from Rhode Island and New York, and Devons from New Jersey. Those animals provided North Carolina farmers with visual evidence

Elisha Mitchell was one of the state's earliest and most ardent advocates of scientific agriculture. Mitchell, a native of Connecticut and a Yale graduate, came to the University of North Carolina in 1818 and remained there as a professor of chemistry, mineralogy, and geology for the next thirty-two years. Photo courtesy of A&H.

To a surprising degree, the early fair devoted a great deal of its effort toward promoting better livestock. Although tobacco and cotton are often regarded as the major agricultural commodities of the nineteenth century, livestock production was crucial to the state's economy. The diet of most North Carolina families relied heavily upon livestock grown on the family farm, and horses and mules furnished the motive power to operate both agricultural equipment and means of transportation, even in cities. North Carolina farmers, whether they lived on small family farms or ran large plantations, depended upon livestock for their livelihood. Leaders of the State Agricultural Society clearly understood this and, as a result, sometimes as much as one-half of the premiums awarded at the state fair in a given year went to livestock exhibits.

Perhaps the fair's most significant achievement in the area of agricultural reform was its introduction of purebred cattle to North Carolina farmers. As late as 1840 there were no purebred cattle in the state, with the possible exception of a few unregistered Shorthorns

of the superiority of blooded cattle over the vast majority of their native animals.

By 1878, several leaders of the society introduced Jerseys to their farms, among them Rufus S. Tucker, Wake County merchant, planter, and railroad promoter; William Grimes, Pitt County planter; Lynn Banks Holt; and William G. Upchurch, all members of the state's agricultural elite. These men began to exhibit their Jersey herds, and the fair responded by placing an increased emphasis on registered cattle. By 1880 animals with pedigrees received preference over those without, and in 1882 the fair mandated that all bulls exhibited have a pedigree, except those in classes of natives and grades, which were reserved for entries of non-blooded cattle that composed the herds of most North Carolina farmers. By 1885 the society required a pedigree of both bulls and cows exhibited in all classes except natives and grades. By adopting the requirement that potential exhibitors produce a pedigree for exhibits of cattle within any of the breed categories, the fair encouraged North Carolina farmers to develop purebred herds, while at the same time demonstrating to them through

cattlemen introduced purebred Ayrshires to the state in 1878, Guernseys in 1882, and Holsteins in 1884. Soon thereafter, those breeds—in all cases the property of State Agricultural Society leaders— appeared at the state fair: Ayrshires and Guernseys in 1885 and Holsteins in 1886. After 1886 the Wake County Cattle Club, many of whose members were leaders of the society, regularly entered excellent blooded-cattle exhibits, as did the state's larger livestock farms, such as the Biltmore Estate and Julian S. Carr's Occoneechee Farm. The fair also continued to feature exhibits of blooded cattle from farms in other states. Herefords appeared at the 1892 fair, decades before there were any Hereford breeders in North Carolina. Additional breeds introduced at the fair by out-of-state breeders included a Brown Swiss bull from Pennsylvania in 1894 and, that same year, Black Angus cattle and a Holstein-Friesian bull that took first prize at the 1893 Chicago World's Fair. Exhibits of purebred cattle from other states played a major role in the development of North Carolina's purebred cattle industry, as herds of registered breeds could be found in

Many of the wealthy members of the North Carolina State Agricultural Society exhibited blooded livestock at state fairs of the late nineteenth century. Among them was Lynn Banks Holt, brother of textile industrialist Thomas M. Holt. Like many of the state's most prominent businessmen, Banks Holt retained an interest in agriculture. Photo courtesy of A&H.

Exhibits of purebred cattle from other states played a major role in the development of North Carolina's purebred cattle industry, as herds of registered breeds could be found in the state soon after they were displayed at the fair. Breeds such as Guernseys, Holsteins, and Ayrshires appeared on exhibit at the fair within two to seven years of being introduced into the state by leaders of the Agricultural Society.

purebred exhibits the obvious advantages of doing so.

For the remainder of the century, the fair served as the premier showcase for the blooded livestock of North Carolina's more progressive farmers. Leading North Carolina

the state soon after they were displayed at the fair. Breeds such as Guernseys, Holsteins, and Ayrshires appeared on exhibit at the fair within two to seven years of being introduced into the state by leaders of the Agricultural Society.

Swine exhibits at the antebellum and early post-Civil War fairs followed the pattern established for cattle exhibits, with animals usually exhibited as small breeds, large breeds, and natives. Exhibits featured swine that were mixtures of the Suffolk, Berkshire, and Chester breeds, but no purebreds were shown. The fair required no pedigree of animals in any class, and only animals judged "worthy" on the basis of their appearance received premiums. Fewer premiums were awarded for swine exhibits than for cattle exhibits. In 1873 the fair for the first time offered premiums by

upgrade their swineherds, and thousands of farmers who might otherwise never have seen these new breeds quickly became acquainted with them.

Although the fair's efforts to improve the state's cattle and swineherds were far more significant than those to improve horses and mules, the fair's management emphasized the latter two animals for years. That emphasis, particularly strong during the antebellum period, began to decline in the early 1880s as cattle exhibits began receiving increased attention. The fair exhibited horses and mules

Through the use of experiments, essays, and lectures, fair officials attempted to encourage North Carolina's farmers to better care for their swine and to abandon the old "root, hog, or die" system of allowing the animals to run at large. Through its exhibits of registered breeds, it encouraged farmers to upgrade their swineherds, and thousands of farmers who might otherwise never have seen these new breeds quickly became acquainted with them.

breed instead of class; examples of breed categories for swine included Essex, Berkshire, Chester, Whites, and Suffolk, although pedigrees were not required, and fair officials assigned animals to exhibits based only on their appearance.

As with cattle, the fair introduced new breeds of swine to the state. In 1876, six years before Berkshire herds appeared in the state, the fair exhibited a pedigree Berkshire from Pennsylvania. The Jersey Red breed of hog, introduced at the 1883 fair, was quickly adopted by North Carolina farmers, and by 1897 the fair offered nine premiums for that breed alone. In the 1880s the state's more progressive farmers began to introduce new breeds of swine, including Poland Chinas, Victorias, and Duroc-Jerseys, which soon appeared at the fair. The fair exhibited some registered swine in the 1880s, but fair officials did not mandate pedigrees in connection with such exhibits until 1894. Through the use of experiments, essays, and lectures, fair officials attempted to encourage North Carolina's farmers to better care for their swine and to abandon the old "root, hog, or die" system of allowing the animals to run at large. Through its exhibits of registered breeds, it encouraged farmers to

in the classes of thoroughbreds, heavy draft horses, light draft and saddle horses, mules, and jacks and jennets. The fair's management continued to slight draft animals (those animals used to pull plows and wagons) with this system for the remainder of the century.

Given the wealth and social status of the Agricultural Society's leadership, thoroughbreds, unsurprisingly, received the most emphasis. The fair required a pedigree for thoroughbreds in 1855, nearly thirty years before such documentation was required for any other animal. Premiums for thoroughbreds were significantly higher, ranging from a few dollars above to more than double those offered for other classes and branches of livestock. Thoroughbreds always led the parade of horses, which began in the antebellum period.

Next to the thoroughbred, the racehorse received the most attention from fair officials. The fair offered a handsome purse to the winners of its annual horse races as a means of encouraging the development of the racehorse, with emphasis upon trotters. The fair also usually staged one or two racing events open only to North Carolina horses as an additional incentive to the state's horsemen to improve the bloodlines of their animals. The development of purebred stocks of racehorses occurred much more slowly than was the case with thoroughbreds, however, and the racehorse was not required to have a pedigree to enter races at the fair until 1885. Very likely the fair had little impact on the improvement of thoroughbred and racing strains in the state, for both types were raised primarily by the relatively few wealthy farmers and were used only for sport or for extremely light work. The fair emphasized these horses because Agricultural Society leaders were among the state's wealthiest citizens and had a long history of participation in the breeding and racing of horses. The average small farmer needed draft animals, however, not status symbols or playthings, and the fair's leadership erred in not giving draft animals more attention.

While the fair paid less attention to draft horses, mules, and jacks and jennets, it did not ignore them completely. It offered much lower premiums within those classes than for thoroughbreds, and it required no pedigree for draft horses until 1885. In addition, fair officials made no effort to exhibit new draft breeds from outside the state. The fair did, however, sponsor plowing matches both for horses and mules, which provided the farmer an opportunity to observe the comparative abilities of draft animals. In addition, those contests enabled "average" North Carolina farmers to measure the strength and endurance of their draft animals to pull their plows and wagons against that of the animals of some of the state's noted agriculturists. After the Civil War, the fair continued to exhibit draft horses and to stage draft-horse pulling contests, although the growing popularity of the mule led to a rapid decline in the use of draft horses by North Carolina farmers. Yet except for awarding premiums for the best jacks and mules, the fair paid little attention to the mule, an animal universally admired for its stamina in the field but totally lacking in glamour.

Sheep, too, received little attention from the fair. The monetary value of premiums offered for the animals were usually equal to, or a bit lower than, those offered for swine. Until 1873, sheep were usually entered in the following classes: imported sheep, natives and mixed-blooded sheep, and fine-wool or middle-wool sheep. Afterward the fair established exhibits by breeds, among them

Merinos, Southdowns, Cotswolds, Oxfords, and Leicesters. Not until 1882 did the fair require pedigrees for sheep exhibited, and then only for rams. With few exceptions, the fair exhibited sheep raised in North Carolina and thus failed to introduce new breeds to the state.

Poultry exhibits at the antebellum and early post-Civil War fairs were meager but began to improve tremendously during the 1880s. By the 1890s, large poultry farms from North Carolina and other states were entering excellent poultry exhibits. At the 1898 fair, premiums were offered for 46 types of chickens, 3 varieties of turkeys, 10 types of ducks, 6 categories of geese, and 10 types of pigeons. Throughout the nineteenth century, the fair remained the best place for the average farmer to become acquainted with the better types of poultry raised in North Carolina and other states. The fair encouraged new methods of poultry production through exhibits by out-of-state poultry farms that featured the use of incubators. North Carolinians, however, preferred the traditional methods of raising fowl and failed to adopt the use of incubators—exhibited as early as 1880 but still regarded as novelties as late as 1895.

Efforts to improve methods of field-crop production comprised the second major emphasis of the fair's program of agricultural promotion, a formidable task because North Carolina was far behind most other states in the application of modern techniques in that regard. Only a few progressive individuals practiced scientific methods of crop production, which many North Carolina farmers regarded as silly and impracticable. The fair initially relied heavily upon the annual address to familiarize farmers with the advantages of scientific farming methods. The state's leading gentlemen farmers and other

Medal awarded at the inaugural fair of 1853, possibly for excellence in field-crop production. Photo courtesy of A&H.

prominent North Carolina agriculturists customarily delivered the addresses, among them Kenneth Rayner, Thomas Ruffin, Elisha Mitchell, and John L. Bridges. Attendance at the fair on the day of the address was usually the largest of the week. Speakers, attempting to overcome North Carolina's agricultural backwardness, promoted ditching, manuring, the planting of cover crops, the use of marl and ashes, and the practice of crop rotation to lift the state out of its agricultural doldrums. Additional topics discussed included the need for agricultural education and the importance of agricultural chemistry. The annual address lost much of its agricultural value during the 1870s when the State Agricultural Society inaugurated the custom of allowing the governor to deliver the address, transforming it into a largely political and social event.

Although meetings of the Agricultural Society attracted fewer people, they promoted agricultural reforms more effectively than did the annual address. Beginning with the initial fair, the society held its meetings each night

During the post-Civil War years, increased social activity in Raleigh during fair week lured many away from the society's nightly meetings, diminishing their educational effectiveness. As before the war, the meetings appealed primarily to the more educated and affluent gentleman farmer rather than to his smaller, less-affluent counterpart, who was most in need of the information imparted. The meetings did, however, enable the state's more progressive farmers to impart new information to a limited number of average North Carolina farmers.

The fair employed a variety of premiums in its efforts to promote scientific agriculture. During the antebellum and early post-Civil War years, the fair awarded premiums for the best essay on each of several agricultural subjects—such as improving worn-out land, proper drainage, crop rotation, preparation of manure, and the use of the pea as a green manure—but abandoned the practice by 1877. To encourage the application of the scientific method to farming, the fair also

> *By the 1880s, agricultural companies, rather than the fair, were awarding . . . premiums. Fertilizer manufacturers and dealers, for example, offered them for the best crops grown with their products. That change reflected a larger, more significant shift toward a highly commercialized agricultural economy in North Carolina and elsewhere in the South, in which manufacturers of agricultural implements, fertilizers, agricultural chemicals, and other farm-related items urged farmers to use their products to increase farm productivity.*

during fair week, after the fair had closed for the day. At those meetings, the society attempted to instruct farmers in the use of scientific agricultural methods—the results of which fairgoers had seen on exhibit during the day. The meetings relied upon speeches, at times accompanied by reports on agricultural experiments. In the antebellum period, leading farmers, often society members, served as speakers, but during the late 1880s professional agriculturists, either college professors or employees of the state Department of Agriculture, spoke on such topics as the importance of irrigation, the best mode of curing tobacco, the farmer's need for education, and chemistry in agriculture.

awarded premiums for certain types of agricultural experiments, among them the costs and results of subsoil plowing, the use of peaty soils as manure, the benefits of guano, and various methods of cultivating rye and tobacco. Just as with premiums offered for essays, however, fair officials discontinued the use of premiums for experiments by 1877. By the 1880s, agricultural companies, rather than the fair, were awarding such premiums. Fertilizer manufacturers and dealers, for example, offered them for the best crops grown with their products. That change reflected a larger, more significant shift toward a highly commercialized agricultural economy in North Carolina and elsewhere in the South, in which manufacturers of agricultural implements, fertilizers, agricultural chemicals, and other farm-related items urged farmers to use their products to increase farm productivity.

The North Carolina Department of Agriculture played a significant role in that process, as well as in the state fair's broader efforts to promote improved agricultural practices among the state's farmers. Immediately following its creation in 1877, the department began to mount scientific agricultural exhibits at the fair. The Agricultural Experiment Station, an agency of the department, soon began to exhibit its laboratory work on fertilizer analysis. After 1887, when the Experiment Station was transferred from the University of North Carolina to the newly created North Carolina College of Agriculture and Mechanical Arts (A&M College), it began conducting actual agricultural experiments, rather than merely serving as an agricultural chemistry lab, and those efforts resulted in the production of clearly superior crops. When those crops were exhibited at the state fair, average farmers could see with their own eyes the advantages resulting from scientific farming methods. Moreover, employees of the Experiment Station were on hand at the fair to instruct farmers in the application of those methods. Farmers appreciated the instructions, some taking notes on the methods explained. The Experiment Station's presentations, offered during the day in an informal and easily imparted manner, did not compete with the many social attractions of fair week and reached a large number of farmers.

During the 1880s, farmers in North Carolina and the rest of the South, traditionally loyal to the Democratic Party, increasingly expressed their belief that the party's leadership was unresponsive to their needs and concerns. The increasingly market-oriented nature of farming prompted the agrarian unrest. While most North Carolina farmers in the post-Civil War era were not exclusively subsistence farmers and relied upon the marketing of crops and livestock for a portion of their income, by the 1880s economic forces were compelling them to invest ever more heavily in the production of cash crops. Those forces included rapidly developing nationwide transportation, communications, marketing, and financial systems, all of which were completely beyond the control of the individual farmer. Pressured by the new economic forces and convinced that incumbent Democrats, their traditional allies, served only the interests of the developing railroad, banking, and telegraph industries, grass-roots farmers began to demand that the party and the state and national governments address their economic concerns.

Leonidas L. Polk, North Carolina's first commissioner of agriculture, in 1886 founded the *Progressive Farmer*, an agricultural journal that became the voice of the state's growing host of politically disenchanted farmers, and led an agrarian rebellion in North Carolina. The following year Polk formed the North Carolina Farmers' Association, which in 1888 merged with the politically powerful National Farmers' Alliance and helped many of its supporters win election to the state legislature. Eventually thousands of rural and small-town North Carolinians became convinced that the Democratic Party would never address their concerns. Under the leadership of Polk and Marion Butler of Sampson County, they bolted the party in 1892 to join with other southern and

Leonidas L. Polk, noted advocate of agricultural reform, became North Carolina's first commissioner of agriculture in 1877. Quickly realizing the potential of the state fair to inform the state's citizens about the department's activities, Polk established a pattern of partnership with the fair. Photo courtesy of A&H.

Marion Butler of Sampson County, agricultural reformer and a leader of the Populist Party in North Carolina, was elected United States senator in 1895. From an engraving in his newspaper, the *Caucasian*, 1895.

western agrarians to form the Populist Party. For a variety of reasons, however, the most important being the perceived need to maintain white supremacy, the majority of North Carolina's farmers remained with the Democratic Party, through which they continued to work for state and national governmental programs that would ease their economic plight. The agrarian revolt had run its course by the end of the 1890s, but not without instigating some important changes.

One of the most significant demands of the agrarian reformers was for the state to provide "practical" agricultural education to its citizenry, and the fair provided the state with an opportunity to do so. Bowing to that demand, in 1887 the state legislature required the Department of Agriculture to hold institutes in each county to promote agricultural reform. During the late 1880s and 1890s the department also sponsored farmer's institutes at the state fair in order to reach the large numbers in attendance. Frequently employing experiments and demonstrations, the institutes addressed such topics as the silo and ensilage, fertilizers, and fairs as a means of promoting agriculture. Soon afterward, A&M College began to enter exhibits at the fair. In addition to the state's use of fair exhibits, private agricultural organizations worked with the fair to promote scientific agricultural methods. The North Carolina State Grange, for example, offered premiums for the best agricultural displays made by one of its members, and during the 1880s and 1890s county societies of the North Carolina Farmers' Alliance, the most

powerful agricultural organization of the nineteenth century, encouraged members to adopt better farming practices by exhibiting the best products raised by Alliance farmers using scientific methods. The *Progressive Farmer*, the journal of the Alliance, suggested that the fair should set aside several of its exhibition halls for the use of the organization and increase competition for premiums between county Alliances, thus stimulating agricultural improvement in the state.

Both the Alliance and the Grange rendered their greatest service to the fair by encouraging their members to attend and support it. Once other institutions capable of actually instructing the state's farmers in new methods of scientific agriculture had been created, it became far more important for the fair to create an interest in and a desire for agricultural reforms. The exhibits, speeches, experiments, and demonstrations seen and heard by farmers encouraged to attend the fair by both organizations helped the fair to fulfill that task.

Since improved farming methods resulted in better yields from each acre of land under cultivation, the fair relied upon per-acre production as the basis on which it offered premiums. The first fair offered premiums for the best per-acre production of some twenty-eight crops, among them cotton, tobacco, and a variety of small grains. That emphasis continued throughout the antebellum period. By 1860 the fair required field-crop exhibits to be accompanied by affidavits stating the amount produced per acre and established per-acre production requirements for each crop

The initial building on the campus of the North Carolina College of Agriculture and Mechanical Arts, founded in 1887, from a 1905 photo. Photo courtesy of A&H.

exhibited. When the fair was reestablished in 1869, the quality as well as the quantity of crops produced received greater emphasis as premiums were also offered for the best exhibits of cotton, tobacco, and small grains, regardless of per-acre production. During the remainder of the century, the fair offered premiums both for quantity and quality of crops produced but generally encouraged larger per-acre yields.

After the Civil War, the fair employed premiums of varying value to encourage the production of certain types of crops over others. Clearly favored were wheat, cotton, and tobacco, which brought premiums as high as fifty dollars per crop—approximately five times the premium offered since the antebellum period for other crops. That differential in premium reflected the North Carolina farmer's gradual shift from subsistence to commercial farming and an increasing reliance upon the production of cash crops. After the Civil War, premiums offered for tobacco remained higher than those offered for other crops, with the exception of cotton, although not as disproportionately higher as they had been in the late antebellum period. Fertilizer companies, too, offered premiums for the best per-acre production of tobacco achieved with the use of their products. In the 1870s, seeking to diversify the state's tobacco crop, the fair unsuccessfully encouraged the

production of a light-colored tobacco to be used as cigar wrappers, offering premiums for that variety five times higher than those offered for regular tobacco strains.

As would be expected in a state that had been a part of the South's cotton kingdom, the fair continued to stress cotton production well after the Civil War. Fertilizer companies also offered premiums for cotton exhibits, usually placing heavy emphasis upon production. Premiums offered for cotton by the fair itself, however, often recognized quality rather than quantity.

While the fair's emphasis upon the increased production of tobacco and cotton is hardly surprising, its efforts to encourage the production of small grains are a bit unexpected. Leaders of the State Agricultural Society in the antebellum period expressed extreme interest in small-grain production, and the fair offered premiums for wheat production equal to those offered for tobacco and cotton. Antebellum and immediate post-Civil War fairs also awarded premiums for the best essay on the production of rye as a food for livestock. After the war, premiums offered for small grains were usually based on quality, not quantity. The fair exhibited various varieties of small grains grown by the state's leading farmers, among them several types of wheat, buckwheat, rye, oats, and barley, as well as newer varieties, especially those that were rust-resistant. Fair premiums also encouraged the growing of grasses, clover, and other hay-producing crops. The fair encouraged the

initially placed less emphasis upon household self-sufficiency and used premiums only to encourage the farmer's production of such items—premiums significantly less than those offered for livestock and field crops. By 1896, however, the attitude of the Agricultural Society's leaders had undergone a decided shift. In its efforts to encourage self-sufficient farming and discourage the state's growing dependence upon an agricultural system dominated by two cash crops—cotton and tobacco—the fair offered a special seventy-five-dollar premium for the best individual display of agricultural products, garden vegetables, fruits, and home industries.

Such efforts had limited impact, however, for they reflected the ideals of the society's more affluent leaders rather than the economic realities faced by the average farmer. Even if the average farmer had desired to engage in diversified farming, the increasingly market-driven agricultural economy, with its

In its efforts to encourage self-sufficient farming and discourage the state's growing dependence upon an agricultural system dominated by two cash crops—cotton and tobacco—the fair offered a special seventy-five-dollar premium for the best individual display of agricultural products, garden vegetables, fruits, and home industries. Such efforts had limited impact, however, for they reflected the ideals of the society's more affluent leaders rather than the economic realities faced by the average farmer.

production of such crops for three reasons. First, many of the fair's antebellum leaders sought plants that could be used as green manure, or cover crops. They found that certain grasses, especially clovers, produced excellent cover and hoped that the state's farmers would adopt the practice of using them. Second, improved livestock herds, on which the fair expended so much effort, required improved feeds, and better varieties of grains, clover, and grasses could fulfill that need. Third, such production promoted the Agricultural Society's ideal of a more diversified agricultural economy.

The fair's enthusiasm for diversified farming resulted directly from its very first efforts, beginning in 1853, to encourage self-sufficiency among the state's farmers by offering premiums for vegetables, fruits, cured meats, and household manufactures. The fair

reliance upon the post-Civil War crop-lien credit system—under which the farmer's highly marketable nonperishable tobacco and cotton were pledged as collateral for the funds borrowed to produce them—would have prevented him from doing so. That economic quagmire, not of their making, ensnared the farmers of North Carolina and the South well into the twentieth century. The crop-lien system eventually yielded only to the cataclysmic events of the New Deal and the Second World War.

Leaders of the State Agricultural Society more than anything else desired to make the fair a family event and thus did not ignore the farm woman. Categories open to female exhibitors reflected, however, the Victorian-era belief that a woman's place was in the home and, preferably, in the kitchen. Ladies entered their handiwork for premiums in several

classes, including preserves and pickles,
canned fruits and vegetables, baked goods,
needlework, embroidery, and artwork. The
relatively low premiums offered for those
goods, usually only one to three dollars,
reflected the comparative value that a male-
dominated society placed upon the efforts

required to produce them. Nonetheless, the
exhibits of such domestic products enabled
the entire farm family to participate in the
competition.

The fair achieved some success in its efforts to encourage farmers to use agricultural
machinery, a major goal. It encouraged

development of new machinery and improvement of the old, functioned as a testing ground for farm implements, and served as a market place for those wishing to buy or sell agricultural implements. It showcased the new and best in agricultural implements from manufacturers in and out of the state, displaying superior quality farm machinery to thousands of North Carolina farmers.

Following its customary practice, the nineteenth-century fair offered premiums in three basic classes of agricultural implements, with several subdivisions within each class. Implements for tilling the soil, such as plows, cultivators, and harrows, composed the first class; farm vehicles such as wagons and dump carts made up the second; and harvesting and processing machinery, such as corn shellers, cotton gins, threshing machines, broadcasting machines, hay presses, stalk cutters, and other implements, composed the third.

Exhibits by out-of-state firms dominated all classes of agricultural implements during fairs of the antebellum period and received little native competition after the Civil War. Firms from Maryland, Virginia, Pennsylvania, New York, New Jersey, and Mississippi exhibited products at the antebellum fairs. After the Civil War, firms from other states, including Georgia, Illinois, Indiana, Wisconsin, Connecticut, and Ohio, participated. The big names in the industry, such as John Deere

Above: Baked goods, too, were considered a category in which women exhibited. This array of baked goods was displayed at the 1985 fair. Photo courtesy of NCDA&CS.

Right: Blue-ribbon-winning baked goods at the fair, ca. 1988. Photo courtesy of NCDA&CS.

L. L. POLK & CO.,
NORTH CAROLINA HEADQUARTERS
FOR ALL KINDS OF

Machinery and Farm Implements,

Engines, Gins, Presses, Cotton Cleaners, Saw Mills, Corn Mills, Cider Mills and everything needed by our people.

Be sure to visit our Exhibit on the Grounds during Fair Week.

Write for Circulars and Prices before buying.

Address   L. L. POLK & CO.,
RALEIGH, N. C.

and McCormick, began to exhibit their products during the 1870s, and by the end of the decade many of them also sponsored premiums for crops, a practice that continued for the remainder of the century.

Implements manufactured in North Carolina were exhibited at the antebellum fairs, but they were always heavily outnumbered by those from other states. Not until after the Civil War did implements made in North Carolina offer any real competition to those made outside the state. For a brief period after the fair was reestablished in 1869, implements manufactured in North Carolina dominated all exhibit classes, but only because few northern firms chose to exhibit their products. By the mid-1870s, however, items from other states had again gained ascendancy. That trend continued as additional agricultural implement dealers became established in North Carolina and employed the fair to display their merchandise, most of it manufactured in other states. The inability of North Carolina firms to produce machines that increased agricultural productivity underscored the essentially colonial relationship North Carolina and the South had with the rest of the nation. North Carolina farmers continued to produce their crops with implements and machinery *made* in other states, but at least the fair allowed them to *see* the best agricultural implements available. The fair also provided a testing ground for agricultural implements entered as exhibits, enabling farmers to weigh their comparative performance capabilities. Fair judges recommended to the farmers of the state those items that proved to be of particular merit.

Exhibits of agricultural implements at the fair also forecast the future of farming. Although most of the implements exhibited were typical of what one would expect at a nineteenth-century agricultural fair, prototypes of implements not perfected until the twentieth century could also be seen. The 1853 fair offered a premium for the best portable steam engine "applicable to agricultural purposes generally, as a substitute for horse power," and steam engines were displayed at the fairs throughout the nineteenth century, although all were

Fruit presses on display at the 1890 state fair. Drawing from the Raleigh *Daily State Chronicle.*

at the fair. The farmer, too, marketed his products exhibited at the fair, but under severe restrictions. The Agricultural Society forbade farmers from selling their exhibits during the fair but allowed them to "book orders" for later delivery. Even then, exhibitors were forbidden to engage in any "loud or noisy" efforts to call attention to their products. Early fair management did allow farmers to sell their exhibits of farm produce, livestock, or other products at auction on the last day of the fair, after the exhibits were closed. After 1880 the fair increased the number of days on which auctions could be held, hoping to make the fair an important market of farm produce, but the other restrictions on sales remained. Despite a gradually increasing emphasis on marketing rather than exhibition, such hopes never materialized, and the nineteenth-century fair never became an important farm market.

The state fair's overall impact upon nineteenth-century agriculture is difficult to measure because the fair represented only one part of the state's agricultural reform movement. Additional factors encouraging reform during the period included many of the state's leading farmers, county and area fairs, the few agricultural journals published in the state, private agricultural organizations, and especially state agricultural institutions such as the Department of Agriculture and A&M College. Their combined efforts gradually transformed agricultural practices in the state, but it is impossible to assign to each an exact fraction of the net result. It is, however, possible to identify some of the fair's contributions to the general reform movement.

Led by some of the state's wealthiest and most progressive citizens, the fair remained in the vanguard of the reform movement. It

stationary. Potato diggers, a "cow milking machine" from England, and a cotton picker from Scotland Neck were among other visionary exhibits encouraged by premiums offered for the best agricultural invention.

Such futuristic exhibits resulted in little practical reform, and the fair's efforts to foster the invention of new agricultural machinery must be judged a failure. Portable steam engines were seldom put to agricultural uses, and the milking machine and the cotton picker were years ahead of their time. Although each year the fair awarded a premium for the best new agricultural invention, usually to some type of plow, no truly significant agricultural implement or machine resulted from the practice.

Although the fair emphasized the educational nature of its exhibits, some aspects of commerce intermingled with its efforts to promote agriculture. This was most obvious in the efforts of manufacturers and dealers to market their fertilizer, guano, farm implements, and other products through exhibits

The Agricultural Society forbade farmers from selling their exhibits during the fair but allowed them to "book orders" for later delivery. Even then, exhibitors were forbidden to engage in any "loud or noisy" efforts to call attention to their products.

Left: While farmers exhibited various crops for premiums, farm women entered a variety of home-canned vegetables, fruits, and other items in competitions for ribbons and prizes. This judging of canned goods took place at the fair around 1988. Photo courtesy of NCDA&CS.

Below: Canned goods with ribbons, ca. 1985. Photo courtesy of NCDA&CS.

introduced blooded cattle to the state because its leaders improved their cattle herds. It encouraged diversified farming because its leaders both favored and practiced doing so. It encouraged the use of cover crops and commercial fertilizers because its leaders used them. And it promoted the use of agricultural machinery because its leaders favored mechanized farming. In short, the fair encouraged North Carolina farmers to adopt the methods and practices its leaders advocated and found successful.

The fair's success at promoting agricultural reform, of course, depended upon its ability to reach farmers from all parts of the state. In fact, only a fraction of the state's farmers attended the nineteenth-century fair, and many of those who did came from Wake and its neighboring counties or, at least, from the eastern part of the state. Throughout the nineteenth century, the press of the western section of the state blamed the Agricultural Society's leaders for ignoring the region, making western farmers reluctant to patronize the fair. Easterners, the western press charged, controlled the fair's leadership and showed marked favoritism to eastern exhibitors. That sectional animosity, something the fair sought

of the more numerous small farmers, as the emphasis placed upon horse racing and thoroughbred horses illustrates. Fair leaders also unintentionally damaged the fair by dominating the competition for premiums, although that domination allowed the average farmer to see superior exhibits. The elitist hegemony received reinforcement in the 1880s and 1890s when such large farms as Biltmore and Julian S. Carr's Occoneechee began to exhibit, to some degree discouraging small farmers from competing. By 1897 the farm journals were suggesting that the situation must be remedied if the fair were to survive.

> *Throughout the nineteenth century, the press of the western section of the state blamed the Agricultural Society's leaders for ignoring the region, making western farmers reluctant to patronize the fair. Easterners, the western press charged, controlled the fair's leadership and showed marked favoritism to eastern exhibitors.*

to overcome, led members of the State Agricultural Society to plead continually for support from farmers throughout the state.

While western Carolinians had reason to complain, a surprisingly large number of North Carolinians passed through the fair's gates. In the nineteenth century, fairs ran for only five days but had an average daily attendance of at least 6,000 persons, or 30,000 attendants per fair. Inasmuch as thirty-seven fairs were staged between 1853 and 1899, approximately 1,110,000 persons attended them. These figures are inflated because numerous individuals attended the fair more than one time, but they nevertheless represent an impressive achievement, especially when one considers the fact that in 1860 North Carolina had a population of only 992,622, and in 1900, 1,893,810. Those attending came from all sections of the state and from all classes of the state's social structure. The exhibits of livestock, field crops, farm implements, and industrial products that they saw certainly contributed to the improvement of agriculture within the state, although of course it is impossible to determine the fair's precise influence upon the state's economic growth in either the agricultural or industrial sector.

The fair's elitist leadership set in motion policies that at times countered the interests

Moreover, fair leaders failed to cooperate as fully as they should have with agricultural organizations such as the Grange and the Farmers' Alliance. But perhaps their worst offense was to look outside the state for reform.

For a variety of reasons, the fair's significance as a factor in the agricultural reform movement declined toward the end of the century. During the antebellum years, few of the state's average farmers read any of the handful of agricultural journals then being published. Since no state-supported agricultural institutions existed, a means of acquainting the state's farmers with the rapid advances in scientific agriculture was sorely needed. The fair did much to meet that need. After the Civil War, by purely quantitative measures, the fair's contributions to the state's agricultural development were greater than before, but its overall contribution to the total agricultural reform movement actually declined. Because the state Department of Agriculture and A&M College undertook much of the work of instructing farmers in newer agricultural methods, leaders of the fair ceased to be as concerned with such matters. Agricultural implement dealerships established throughout the state served to reduce the significance of that aspect of the fair's work. More than ever, the fair became merely a showplace of the state's agricultural development rather than an instigator of reform. Still, fair exhibits made farmers aware of the state's agricultural progress, aroused pride in those accomplishments, and promoted hope for continued progress in the future.

Promotion of Agriculture at the Twentieth-Century Fair

For the first quarter of the twentieth century, during which the state fair remained under the control of the North Carolina State Agricultural Society, its efforts at agricultural promotion changed little. Its exhibit structure, for example, remained essentially unchanged from the turn of that century until 1922. Premiums increased in amount, but only gradually and not always at the rate of inflation experienced within the general economy. During that time the fair's premium lists continued to emphasize field crops, especially corn, cotton, tobacco, and small grains—wheat, oats, and barley. Livestock, divided into separate departments for horses, cattle, sheep, swine, and poultry, remained the second most significant exhibit category. (The breeds arrayed within each department, however, were the same as those exhibited at the nineteenth-century fairs.) Other agricultural premium categories during the period were horticulture—including fruits, flowers, and nursery plants—and agricultural machinery and implements.

By 1900 the fair had returned to its policy of emphasizing its educational value by prohibiting the sale of agricultural crops and livestock exhibited there. Regulations governing

the development of a commercial fair, the society did allow exhibitors "to book and receive orders" for delivery after the fair. That outright prohibition on the sale of exhibits changed in 1910 when the fair allowed exhibitors to sell their goods, but again with the stipulation "that no articles on exhibition shall be delivered until after the closing of the fair," a restriction aimed at emphasizing the educational value of all exhibits until the fair closed. The Agricultural Society made no additional changes to that policy while the fair remained under its control.

The fair's premium list of 1910 exemplified the emphasis the fair then placed on certain agricultural activities, but it was also remarkably similar to such lists of the nineteenth-century fair in terms of exhibit categories. Under its lead classification, Department A, field and garden crops, the fair awarded premiums for the cash crops cotton, corn, rice, peanuts, and tobacco, as well as for a variety of grains, including wheat, oats, rye, and barley. Cover crops eligible for premiums included clovers, vetch, soybeans, hay, and grasses. Truck crops considered for premiums included Irish and sweet potatoes, cabbage, onions, celery, turnips, collards, squash, lima beans, cucumbers, peppers, cauliflower, rutabagas, tomatoes, and a miscellaneous category. Exhibitors received awards based on the quality of their product rather than the amount of yield per acre.

Horses, which topped the list of livestock premium categories, included a mixture of draft and sport animals. Breeds and types included Percherons, French and German Coach, Hackneys, Standard Bred Horses,

*B*y 1900 the fair had returned to its policy of emphasizing its educational value by prohibiting the sale of agricultural crops and livestock exhibited there. Regulations governing exhibits in that year warned that "Exhibitors will not be allowed to sell goods during the fair, nor will they be permitted to call attention to their wares in any noisy or disorderly manner."

exhibits in that year warned that "Exhibitors will not be allowed to sell goods during the fair, nor will they be permitted to call attention to their wares in any noisy or disorderly manner." While clearly seeking to discourage

saddle horses, roadsters, Shetland ponies, jacks, jennets, and mules. The fair clearly emphasized working animals, whether they were hitched to a plow or a carriage. Cattle continued to be exhibited without distinction between dairy and beef cattle, and the breeds for which the fair offered premiums—Jerseys, Guernseys, Holsteins, Ayshires, Red Polled Herefords, Herefords, Aberdeen Angus, Shorthorns, and Grades (the last category open only

displays about farm crops, agricultural engineering, and soil chemistry from the agronomy department.

Every year, the fair came to rely upon the Department of Agriculture to mount major

The fair recruited faculty from the college to serve as judges of agricultural exhibits and as directors or co-directors of entire categories of exhibits, and by 1924 North Carolina State College faculty members served as "superintendents" for practically all of the fair's exhibit categories.

to North Carolina breeders)—remained essentially those eligible for premiums at nineteenth-century fairs. Swine categories, too, remained little changed.

The fair's management increasingly integrated personnel from the state's two major agricultural institutions, A&M College (renamed North Carolina State College of Agriculture and Engineering in 1917) and the Department of Agriculture, into the operation of the fair. The location of both the college and the department in the capital city facilitated that integration, and gradually the fair came to rely upon the cooperation and help of officials and staff from both institutions. The fair recruited faculty from the college to serve as judges of agricultural exhibits and as directors or co-directors of entire categories of exhibits, and by 1924 North Carolina State College faculty members served as "superintendents" for practically all of the fair's exhibit categories. Students at State College likewise contributed. In 1921 they began to hold their own fair in conjunction with the state fair (a convenient arrangement inasmuch as the college campus was immediately across Hillsborough Street from the fairgrounds). In 1925 the students moved their fair, which featured an exhibit on broiler chicks in the poultry category, across Hillsborough Street to the fairgrounds. The student fair remained a prominent feature of the fair's agricultural exhibits into the 1940s. Academic departments at State College also exhibited at the fair in 1925 and thereafter. Exhibits in 1925 included the agricultural administration department's prize winner, "A Farm Without Records is Like a Clock Without Hands"; wax models of plants and insects and a "mammoth microscope" from the biology department; and

exhibits designed to instruct fairgoers in modern agricultural practices. At the 1915 fair, for example, the department urged fairgoers to "let the experts of the Department prescribe for your waste land." The department's Veterinary Division demonstrated its efforts to eliminate hog cholera, its Drainage Division's exhibit explained ditch-and-tile field drainage methods, and its Extension Division urged that "Farmers call on the Agricultural Extension Service, Its Purpose is to help you." The Agronomy Division displayed "model" farms for the Mountain, Piedmont, Coastal Plain, and beach-land regions of the state, along with methods for eliminating the "worst" weeds, while the Division of Horticulture and Entomology demonstrated how to care for fruit plants and treat them for disease. Exhibits by the Department of Agriculture emphasized practical solutions to problems faced by the state's farmers, as exemplified by its 1920 exhibit on successful methods for eliminating the cattle tick, a deadly threat to the state's cattle herds that agricultural agencies throughout the South were seeking to control and eliminate.

The fair also continued its efforts to encourage the development of new agricultural implements, but increasingly it became merely a venue at which agricultural implement manufacturers could display their wares to the farmers in attendance. In 1905, for example, the fair offered premiums for the "best" cotton gins, cotton-bailing presses, guano distributors, threshers, cotton planters, reapers and mowers, hand-operated corn shellers, harvesters, plows, harrows, and fencing. By 1915, modern gas engines and mill machines had been added to the list. Field implements continued to rely upon the power of draft animals rather than tractors, although by 1915 the fair was exhibiting portable steam engines and gas engines on trucks, both of which could be used to power agricultural machinery. In 1920 the State Agricultural Society recognized that the premiums it offered for agricultural implements could have little impact upon innovation in what had become one of the nation's major industries. It announced that because of "the impracticality of establishing a system of judging automobiles,

machinery, mechanical and manufactured exhibits," all such exhibits would no longer be awarded cash premiums. Rather, the fair would offer, "at a nominal rental" to dealers of such machinery, including agricultural implements, exhibit space in which they could display their products. Thus, in the realm of agricultural implements and machinery, the society bowed to the inevitable and acknowledged the fair's obvious limitations as a promoter of innovation, while at the same time

Even before it obtained control of the fair, the North Carolina Department of Agriculture had become a major exhibitor there. At the exhibits shown in these two photos (ca. 1975), the agency explained its role in enforcing weights and measures used to measure agricultural and consumer products. Photos courtesy of NCDA&CS.

reinforcing its valuable role in displaying to North Carolina farmers some of the best agricultural implements. Even that role, however, depended upon the ability of the fair to draw an audience of farmers large enough to entice the state's dealers in agricultural implements to rent exhibit space.

Fortunately, it did so, as evidenced by the exhibit of the International Harvester Company at the 1925 fair, which included a variety of plows, drills, and harrows, as well as a motorized reaper-binder. The featured item in the company's exhibit, however, was a Farmall tractor, which, according to a Raleigh reporter, was "the first one ever brought into North Carolina, although it has been used in the west for a couple of years." The tractor, demonstrated with a variety of plows, cultivators, and mowing attachments, signaled the future for North Carolina farmers. "It has been claimed," the reporter noted, "that this machine can do all the work on a farm that a mule can do." He might have added that the tractor could do a mule's work more efficiently, more quickly, and less expensively, in addition to requiring much less physical

exertion by the farmer who operated it. All that stood between North Carolina farmers and a stampede to buy tractors was the availability of hard cash.

As previously noted, the election of Edith Vanderbilt to the presidency of the State Agricultural Society in 1920 signaled a major break with the society's old leadership and resulted in significant changes in the fair's promotional efforts. An inaugural horse show in 1921, with its emphasis on sport and pleasure animals rather than working farm animals, presaged the major reorganization of the fair's premium categories that occurred in 1922. In that year, the society listed field crops, which had traditionally enjoyed the top spot in premium lists and in terms of total moneys awarded, in the seventh slot. Horse racing, called speed trials, received the top spot in the list, followed by the Horses division, which included the horse show with its emphasis on show animals and equestrian events. Successive slots were devoted to cattle, swine, sheep, and poultry. Clearly, under Vanderbilt, the society had decided to highlight the horse culture as followed by the state's agricultural elite

While Edith Vanderbilt emphasized bloodlines at the fair's horse shows of the 1920s, the mule remained the working animal most valuable to North Carolina farmers. When the state assumed responsibility for the fair, recognition of the mule's privileged status in exhibit categories resumed. Pictured here is a "Mule Pull" at the 1938 fair. Photo courtesy of A&H.

and to relegate field crops to an inferior position within the fair's hierarchy of significant agricultural matters.

The State Agricultural Society may have demonstrated questionable judgment in reordering its priorities, but its new premium structure demonstrated an understanding of the changes that were occurring within the state's farming communities. Most significant was the society's recognition of the new infrastructure that provided information about scientific agriculture and modern farm living to North Carolina's farm families. The 1922 premium list created three new categories through which the society hoped to integrate that infrastructure into the fair.

The first was the Boys and Girls Clubs category. The formation of private clubs to promote improvements in various activities, from gardening to school formation, had been a hallmark of the Progressive movement in the United States, which occurred between the turn of the twentieth century and America's entry into the First World War in 1917. Women often led in the formation of such clubs, a phenomenon that reflected their own successful efforts to obtain the right to vote. Predictably, the formation of self-improvement clubs penetrated into rural America. By far the leading such agricultural group in North Carolina and the South was the 4-H Club, founded in Ahoskie in 1909. Essentially, 4-H Clubs were a rural version of the Boy and Girl Scouts, with a large dose of Progressivism embedded in the organization's motto: "Head, Heart, Hands and Health." Members pledged to apply their heads by obtaining knowledge, their hearts by being considerate of others, their hands by working on specific projects, and their health by observing appropriate dietary and personal hygiene rules. As the 4-H movement swept through schools of North Carolina's eastern counties, clubs began to exhibit at the fair. At the 1915 fair, for example,

girl 4-H Clubs from twenty-nine counties displayed canned goods. The Boys and Girls Club exhibit category extended the fair's official recognition to the 4-H Club movement, which continues to contribute substantially to agricultural exhibits at the modern fair.

Two other exhibit categories—Vocational Agricultural Schools and Home Economics—recognized the role of the federal government in creating an educational infrastructure that reached into the homes of American, and North Carolina, farm families. The administration of President Woodrow Wilson in many ways represents the height of the Progressive movement. Wilson, himself a native southerner and a Democrat, worked with a Congress dominated by southern Democrats, including a number of prominent North Carolinians, anxious to funnel federal government revenues collected through the income tax, which had been established in 1913, into programs that would benefit the South's ailing economy. That strategy worked effectively, particularly in the realm of agriculture.

In 1914 Congress passed the Smith-Lever Act, which provided federal dollars to match state funds to create an agricultural extension program through cooperation between the U.S. Department of Agriculture and the land grant colleges in the various states, one of which was North Carolina State College.

Exhibits by the 4-H Club have been a feature of the state fair since the second decade of the twentieth century. An exhibit of 4-H crafts at a state fair of the mid-1980s. Photo courtesy of NCDA&CS.

Youth sewing exhibits at the fair, ca. 1987. Photo courtesy of NCDA&CS.

Fairgoers at the 1986 fair view a booth designed to appeal to former 4-H Club members. Photo courtesy of NCDA&CS.

Working through the college's agricultural extension division, the program placed extension agents in each of the state's counties. Male agents worked directly with farmers, individually or through local agricultural clubs and organizations, to teach them about the latest developments in scientific farming. Female agents worked with a variety of clubs and organizations to instruct farm women in home economics, which included everything from canning fruits and vegetables to making fashionable clothing and properly decorating the home. In 1917 Congress passed the Smith-Hughes Act, which provided matching federal funds for instruction in agriculture, home economics, and the vocational trades in the high schools. The work of the extension agents and the vocational instructors generated a plethora of clubs and organizations, among them the Future Farmers of America

Above: A North Carolina Extension Homemakers exhibit at the state fair, ca. 1985. Photo courtesy of NCDA&CS.

Right: Members of the Future Farmers of America (FFA), a legacy of the Progressive era, at the 2002 fair. Photo courtesy of the author.

and the Future Homemakers of America, which engaged in a wide range of activities. The state fair provided the ideal venue for displaying the results of such projects and activities, which served to encourage even greater activity by existing organizations and the creation of others. From 1922 until the present, exhibits by such organizations have remained a basic ingredient of the fair.

The 1928 state fair, the first held under the auspices of the Department of Agriculture, as a matter of convenience retained the Agricultural Society's system of classifying exhibits. But it soon began to make changes that transformed that system, with its traditional emphasis on exhibits entered by individual farmers, into essentially a collection of commercial displays developed by a variety of federal and state agricultural agencies. In the following words, the 1928 premium list reflected that shift in emphasis noting that "Here the manufacturer and merchant may display their merchandise for the benefit of themselves and their customers." Fair officials predicted that manufacturers of farm implements and equipment would see the fair as their "best opportunity to place the latest types of machinery designed for efficient service" and would take advantage of the new policy on exhibits. Exhibits such as that of the Raleigh International Company at the 1930

fair suggest that agricultural implement dealers did just that. The company demonstrated a variety of "Modern Farming and Dairy equipment" produced by McCormick-Deering, including a two-row Farmall corn picker, Farmall tractors, and a variety of harvesting machinery.

In recognition of the development of a highly commercialized form of agricultural production and a corresponding expansion of the agricultural bureaucracy within state government, the fair's new management changed its policy concerning sales during the fair. In 1928 fair officials announced their intention to conduct auctions of livestock. Significantly, the fair management conducted the auction in cooperation with personnel from the North Carolina Department of Agriculture. According to the *News and Observer,* North Carolina farmers purchased "a dozen or more animals" at the sale, with prices ranging from $150 to $300. The auction proved "popular with both farmers and breeders since it affords a medium through which seller and buyer may trade." From that time

Winn-Dixie Stores purchased this reserve champion at the fair around 1980. Photo courtesy of NCDA&CS.

Swine, too, remained a feature of the fair's livestock exhibits and markets. This Hampshire boar was a champion at the 1947 fair. Photo courtesy of N&O.

forward, livestock auctions, although they would become far more sophisticated, remained an essential feature of the twentieth-century fair. Indeed, so popular did they become that by the 1960s, retailing and meat-processing firms used livestock purchases at the fair for promotional purposes. In 1966, for example, Winn-Dixie, an operator of grocery stores throughout the South, purchased the fair's grand champion steer for $116 per pound, and Frosty Morn, a meat packer, purchased the swine champion at a price of $30 per hundred pounds. In 1979 McDonald's paid a record $7,500 for the fair's junior steer show champion—a figure that rose to $10,000 by 1983 when Neese's Country Sausage, a Greensboro firm, purchased the swine champion for $4,500 and Winn-Dixie paid $925 for the champion sheep.

In 1929 the state fair completely abandoned the Agricultural Society's old exhibit categories and returned to a structure that resembled the one in effect prior to the drastic

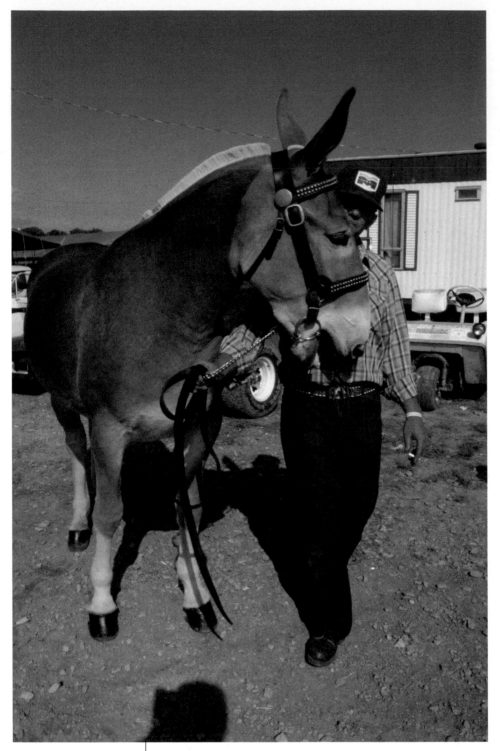

During the years just before and immediately following the Second World War, the tractor replaced the mule in North Carolina's farm fields. Despite its demise as the indispensable draft animal, the mule retains a place in the hearts of rural North Carolinians, as well as a spot on state fair horse-show programs. This mule has been prepared for exhibit at the 1982 fair. Photo courtesy of NCDA&CS.

agricultural and home economics exhibits. With minor revisions, that exhibit structure remained in place until 1952.

During the middle of the twentieth century, and especially in the immediate post-Second World War era, agriculture in North Carolina and the South underwent an unprecedented transformation. The backbreaking, bone-wearying hand labor required of all members of the farm family that had been the hallmark of southern agriculture gave way to mechanized farming. The tenant labor system that had evolved after the Civil War, especially prevalent in the state's cotton-growing regions but also a feature of tobacco cultivation, collapsed as landowners replaced men, women, and children with machines and consolidated small family farms into large holdings. Although the demise of the tenant-farming system and the small family farm and the corresponding growth of agribusiness was a messy, merciless process, it did not exact the heavy human toll historically associated with such major economic dislocations. Fortunately, the hundreds of thousands of workers that agricultural mechanization forced off the farm entered the state's rapidly expanding industrial economy and peopled its burgeoning urban centers, generally improving their economic status in the process.

changes of 1922, which emphasized exhibits by clubs and organizations. Agricultural crops once again reigned, especially cotton, tobacco, and corn, followed by horticultural exhibits, meaning fruits, flowers, and nursery displays. Next followed livestock exhibit categories, with a slot for horses and mules listed after cattle, swine, and poultry. The horses-and-mules category was devoid of any hint of horse shows or equestrian exhibits, instead featuring the lowly draft animals that North Carolina farmers followed behind with their plows. The new structure did, however, retain categories for 4-H Clubs and vocational

Mechanization of North Carolina's farms began with President Franklin Roosevelt's New Deal agricultural programs, which provided landowners with the means to purchase tractors. The country's insatiable appetite for food and fiber during the World War II years accelerated the process. Especially in cotton-growing regions, landowners found that they could replace five mules and five wage or tenant families with a single tractor. The cost to the landholder, underwritten by federal funds, was negligible, usually no more than three to five hundred dollars. In a twenty-county area of North Carolina's Coastal Plain, the number of tractors increased from 1,151 in 1925 to 2,576 in 1940 and to 7, 543 in 1945, while farm units became larger and more diversified. Only the need for a labor force to pick cotton kept the tenant system in place. By 1945, however, an effective mechanical cotton picker was on the market. Although the small size of most North Carolina cotton farms delayed the employment of the device in the field, the International Harvester office in Charlotte shipped 186 of them to farms in the region in 1953. Indeed, by the middle of the decade, the majority of the state's cotton production had been completely mechanized.

Tobacco, the state's most important cash crop, proved more resistant to mechanization. Never as reliant upon tenant labor as cotton cultivation, independent landowning farmers, many of whom cultivated small plots, grew the state's tobacco crop. Because the crop was so lucrative, a farmer, whether tenant or landowner, could support a family on as few as five acres of tobacco. The production of tobacco was also notoriously labor intensive.

Preparing and maintaining seedbeds, planting the crop, and cultivating it, despite the increasing use of tractors, required large amounts of hand labor. During harvesting, curing, and preparation for market, each individual leaf was handled between eight and ten times before reaching the market floor. The process resulted in a rich, aromatic leaf, unsurpassed in quality, and sustained thousands of North Carolina families on small farms.

Technology eventually overwhelmed the small North Carolina tobacco farm. After the Second World War, larger landowners purchased tractors and dismissed tenants, turning increasingly to wage labor. Small landowners, too, began to purchase tractors for the planting and cultivation of their tobacco crops. In 1961 the federal government allowed farmers to lease tobacco acreage allotments, and in 1968 it allowed loose-leaf sales in the entire region's tobacco markets, eliminating the labor-intensive process of hand "grading" each leaf that went to market. Loose-leaf marketing allowed North Carolina farmers to begin to

The enormous changes in North Carolina agricultural production that occurred in the last half of the twentieth century are reflected in these pie charts for 1964 and 1999, prepared for an exhibit on the tenure of Commissioner of Agriculture Jim Graham. The rapid decline in the significance of tobacco, a cash crop in the Tar Heel State for more than 350 years, is striking. Charts courtesy of NCDA&CS.

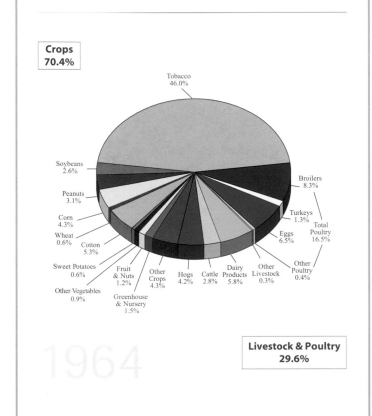

North Carolina Agriculture in 1964
Total Cash Receipts $1.2 Billion

Crops 70.4%

Tobacco 46.0%
Soybeans 2.6%
Peanuts 3.1%
Corn 4.3%
Wheat 0.6%
Cotton 5.3%
Sweet Potatoes 0.6%
Other Vegetables 0.9%
Fruit & Nuts 1.2%
Greenhouse & Nursery 1.5%
Other Crops 4.3%
Hogs 4.2%
Cattle 2.8%
Dairy Products 5.8%
Other Livestock 0.3%
Other Poultry 0.4%
Eggs 6.5%
Total Poultry 16.5%
Turkeys 1.3%
Broilers 8.3%

Livestock & Poultry 29.6%

1964

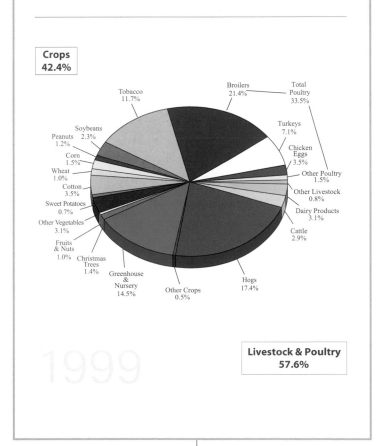

North Carolina Agriculture in 1999
Total Cash Receipts $6.7 Billion

Crops 42.4%

Tobacco 11.7%
Soybeans 2.3%
Peanuts 1.2%
Corn 1.5%
Wheat 1.0%
Cotton 3.5%
Sweet Potatoes 0.7%
Other Vegetables 3.1%
Fruits & Nuts 1.0%
Christmas Trees 1.4%
Greenhouse & Nursery 14.5%
Other Crops 0.5%
Hogs 17.4%
Cattle 2.9%
Dairy Products 3.1%
Other Livestock 0.8%
Other Poultry 1.5%
Chicken Eggs 3.5%
Turkeys 7.1%
Total Poultry 33.5%
Broilers 21.4%

Livestock & Poultry 57.6%

1999

SOMETHING TO CROW ABOUT · 1999

Major NC Farm Commodities

NC RANK	ITEM	1999 CASH RECEIPTS (MILLION DOLLARS)
1	Broilers	1,430
2	Hogs	1,160
3	Greenhouse/Nursery	973
4	Tobacco	784
5	Turkeys	475
6	Cotton	234
7	Chicken Eggs	231
8	Dairy Products	208
9	Cattle & Calves	193
10	Soybeans	154
11	Corn	102
12	Christmas Trees	92
13	Peanuts	82
14	Wheat	68
15	Sweet Potatoes	48
16	Cucumbers	28
17	Apples	26
18	Irish Potatoes	21
19	Tomatoes	21
20	Bell Peppers	15
21	Strawberries	14
22	Blueberries	13

1999 CASH RECEIPTS ALL COMMODITIES $6.7 BILLION

North Carolina's agriculture industry, including food, fiber and forestry, contributes over $48.8 billion annually to the state's economy, accounts for over 22 percent of the state's income, and employs 21 percent of the work force.

How North Carolina Agriculture Compares With Other States

1999 Production

US RANK	ITEM	PRODUCTION		TOP THREE STATES
1	Total Tobacco	449	(Mil Lbs)	NC, Ky, Tn
	Flue-Cured Tobacco	437	(Mil Lbs)	NC, SC, Ga
	Turkeys Raised	46.5	(Mil Hd)	NC, Minn, Mo
	Sweetpotatoes	3,770	(000 Cwt)	NC, La, Calif
2	Hogs & Pigs (12-1-99)	9.5	(Mil Hd)	Iowa, NC, Minn
	Hogs & Pigs Cash Receipts	$1,160	(Mil $)	Iowa, NC, Minn
	Christmas Trees Cash Receipts	$92	(Mil $)	Ore, NC, Mich
	Cucumbers for Pickles	78	(000 Tons)	Mich, NC, Tx
	Trout Sold	4,679	(000 Lbs)	Idaho, NC, Calif
3	Poultry & Egg Products Cash Receipts	$2,238	(MIL $)	Ark, Ga, NC
4	Greenhouse & Nursery Cash Receipts	$973	(Mil $)	Calif, Fla, Tx
	Commercial Broilers	675	(Mil Hd)	Ga, Ark, Ala
	Blueberries	13,000	(000 Lbs)	Mich, NJ, Ore
	Peanuts	299	(Mil Lbs)	Ga, Tx, Ala
	Strawberries	176	(000 Cwt)	Calif, Fla, Ore
5	Burley Tobacco	12.5	(Mil Lbs)	Ky, Tenn, Va
	Net Farm Income	$1,974	(Mil $)	Calif, Tx, Fla
7	Cotton	816	(000 Bales)	Tx, Calif, Ms
	Livestock, Dairy & Poultry Cash Receipts	$3,850	(Mil $)	Tx, Calif, Nebr
	Catfish Sold	2,542	(000 Lbs)	Miss, Ala, Ark
	Apples	170	(Mil Lbs)	Wash, NY, Mich
	Peaches	28.0	(Mil Lbs)	Calif, SC, Ga
	Rye	644	(000 Bu)	N Dak, Ga, Okla
	Watermelons	1,291	(000 Cwt)	Fla, Tx, Calif
8	Tomatoes	713	(000 Cwt)	Fla, Calif, Va
9	Chickens (12-1-99), (Excludes Broilers)	17.1	(Mil Hd)	Ohio, Iowa, Calif
	All Commodities Cash Receipts	$6,687	(Mil $)	Calif, Tx, Iowa
	Crop Cash Receipts	$2,837	(Mil $)	Calif, Fla, Ill
10	Chicken Eggs	2,587	(Mil Eggs)	Iowa, Calif, Pa
11	Sweet Corn	560	(000 Cwt)	Calif, Fla, Ga
12	Grapes	1,900	(Tons)	Calif, Wash, NY
	Export Shares	$1,206	(Mil $)	Calif, Iowa, Nebr
13	Pecans	1,200	(000 Lbs)	Ga, Tx, N Mex
15	Number of Farms	58	(000)	Tx, Mo, Iowa
	Soybeans	29.9	(Mil Bu)	Iowa, Ill, Minn
16	All Irish Potatoes	3,410	(000 Cwt)	Idaho, Wash, Wisc
	Sorghum Grain	552	(000 Bu)	Kans, Tx, Nebr
	Oats	2.0	(Mil Bu)	Wisc, Minn, N Dak
18	Winter Wheat	28.4	(Mil Bu)	Kans, Okla, Tx
	Barley	1.5	(Mil Bu)	N Dak, Mont, Idaho
19	Corn for Grain	51.2	(Mil Bu)	Iowa, Ill, Nebr
31	Cattle on Farms (1-1-00)	940	(000 Hd)	Tx, Nebr, Kans
	Milk	1,216	(Mil Lbs)	Calif, Wisc, NY
33	Hay	1,544	(000 Tons)	Tx, S Dak, Calif

By the end of the twentieth century, animal husbandry, not cash crops, dominated North Carolina's agricultural economy. Agricultural products, ranked by economic significance, 1999. Chart courtesy of NCDA&CS.

In 1999 North Carolina continued to lead the nation in the production of flue-cured tobacco, although tobacco was no longer the state's leading agricultural commodity. Chart courtesy of NCDA&CS.

shift from the old oil- or propane gas-burning flue-curing barns to bulk-curing barns, in which tobacco leaves could be tightly packed. Filter cigarettes, especially flavored filter cigarettes, helped the process, because finely graded tobaccos were less important to their manufacture. Bulk barns also decreased the number of harvestings, or primings, per season, because the precise "ripeness" of the harvested leaves mattered less. In 1972 only 8 percent of North Carolina tobacco farmers used bulk barns; by 1979, 61 percent did so. The mechanical harvester provided the last step in the process of converting the small family tobacco farm, whether tenant or owner operated, into large agribusiness operations. That device, introduced in the early 1970s, harvested more than half of the state's tobacco crop by 1980. The mechanization of tobacco farming resulted in the loss of thousands of small family farms, which were combined into large agribusiness units. During the years from 1964 to 1978, the size of the average

tobacco farm more than doubled, rising from 5.2 to 12.2 acres, while in the decade of the 1970s North Carolina lost a total of 32,546 tobacco farms.

The state fair played an important role in the process by serving as an important means of displaying and demonstrating the latest technological innovations. Especially between the revival of the fair in 1946 and the 1970s, the fair provided dealers in agricultural implements the opportunity to reach thousands of the state's farmers within a short period of time, and major manufacturers availed themselves of that opportunity. At the opening ceremonies of the 1950 fair, for example, Gov. Kerr Scott wandered through what a Raleigh *News and Observer* reporter accurately described as "fields of modern farm machinery" available for a hands-on inspection by thousands of North Carolina farmers. During these years of rapid mechanization on North Carolina's farms, all the major American manufacturers of agricultural implements—International Harvester, John Deere, Ford, Massey-Ferguson, Allis-Chalmers, Oliver, and others—displayed and demonstrated their products at the fair. By 1953, the fair featured a tractor and farm

especially for the small farm. Exhibitors plied interested farmers with promotional brochures touting the performance of their equipment, sought their attention with brightly colored banners, and enticed them

During the 1950s and 1960s, the fair's management recognized the importance of exhibits devoted to agricultural implements through such promotional activities as "Farm Machinery Day," at which implement dealers sponsored 4-H tractor-driving contests in Dorton Arena.

machinery parade on the racetrack in front of the grandstand, but it was at the open-air exhibits by agricultural implement dealers that farmers could examine the machinery up close. Those displays, which by 1965 featured such foreign firms as the English manufacturer David Brown, included tractors of all shapes, sizes, and colors; motorized cotton pickers, corn harvesters, reapers and harvesters; plows and harrows of every size; and mowing, digging, and other equipment operated by a tractor's external power drive. By the late 1960s, manufacturers of bulk tobacco barns were demonstrating their products. By the mid-1970s, several manufacturers were exhibiting mechanical harvesters, including an ill-fated tractor-mounted model designed

to their displays with a variety of promotional gimmicks, including model tractors designed as toys for their children. During the 1950s and 1960s, the fair's management recognized the importance of exhibits devoted to agricultural implements through such promotional activities as "Farm Machinery Day," at which implement dealers sponsored 4-H tractor-driving contests in Dorton Arena.

Paradoxically, the mechanization of North Carolina's farms led to the demise of the fair as a major showcase for agricultural implements. Dealers in agricultural implements could be found in more and more urban centers that served surrounding farmlands, making it easier for farmers to shop for equipment. Improved transportation

During the middle years of the twentieth century, the state fair was a major venue for exhibiting heavy agricultural equipment. These agricultural implements are arrayed in front of grandstand crowds at the 1952 fair. Photo courtesy of NCDA&CS.

By the 1970s the fair had become an outmoded venue for exhibiting heavy farm equipment. The only vestiges of the large displays of the immediate post-World War II years still in evidence at the 2002 fair were a tobacco harvester and a bulk tobacco curing barn, both specialized equipment made by a North Carolina firm. Photos courtesy of Blankinship and the author respectively.

systems, including the nation's best-paved farm-to-market road network and the interstate highway system, removed farm families from their traditional position of relative isolation, providing them easy access to local and regional market centers. Increasing numbers of farm publications carried advertisements for agricultural implements that could be seen anytime at local or regional farm implement dealerships. The fair mattered less and less as a showcase for the latest tractor or combine, the most recent cotton picker or tobacco harvester, the latest bulk tobacco barn curing system. In addition, the fair increasingly appealed to an urban and suburban population, the children of farm families who now resided

in locales such as the Research Triangle Park or in the Piedmont Triad area, who came to the fair to remember their rural heritage and enjoy the midway rides, not to purchase a tractor or new cultivator.

While the fair's collection of antique farm implements expanded to become an important feature of Heritage Circle, what had been acres of space devoted to the display of agricultural implements shrank to practically nothing. By the mid-1960s, the fair's formal program no longer included the Parade of Farm Machinery in front of the grandstand, although exhibits of agricultural implements, by that time known as the Farm Machinery Show, remained a feature. A decade later, even the Farm Machinery Show had been dropped from the official program (although individual dealers continued to offer exhibits). Instead, the fair began to regard farm machinery as a form of entertainment in the form of a tractor pull, which first appeared on the official program of the 1972 fair. In that year, contestants had to be from North Carolina and were limited to tractors with rubber tires. Contest rules prohibited chains, dual tires, and four-wheel drives. In 1973, however, the fair initiated a second class of tractors with "souped-up" engines, which anyone could enter. The tractors pulled a weighted sled.

The tractor pull, at which custom-designed vehicles compete for prizes, remains one of the few vestiges of the once-proud displays of tractors and other agricultural motorized equipment seen at the state fair. In this photo, a slightly modified tractor competes in front of grandstand crowds at the 1982 fair. Photo courtesy of NCDA&CS.

A "tractor" built especially for tractor pulls (and which will never see a day in the fields) springs from the starting line at the 1997 fair. Photo courtesy of NCDA&CS.

The weight shifted forward during the "pull," increasing the degree of difficulty until the tractor was unable to move the sled. The tractor pull, an immediate success, has remained a feature at the fair since its introduction, although the machines now entered hardly resemble the average farm tractor.

In yet another transformation of the state fair's old agricultural implement exhibits, the Southern Farm Show now displays agricultural machinery and implements to farmers and dealers alike. It does so not at the fair itself but at a completely separate event held at the fairgrounds. The farm show is one of several

exhibition. Following completion of the Graham Building in 1976, however, Jim Graham invited Zimmerman to hold the show in Raleigh. The Southern Farm Show, Graham realized, would allow the state fair to regain its status as a major exhibitor of farm machinery, although not within the framework of fair week exhibits. Zimmerman was pleased to accept the commissioner's offer.

The Southern Farm Show, held at the state fairgrounds every February, is billed as the "Largest Agricultural Exposition in the Carolinas and Virginia." It is open only to

The Southern Farm Show, held at the state fairgrounds every February, is billed as the "Largest Agricultural Exposition in the Carolinas and Virginia." It is open only to exhibitors from agricultural and related industries and thus attracts primarily an audience directly involved in the state's agricultural economy.

major trade shows produced by Southern Shows, Inc., whose president, Robert Zimmerman, is a native North Carolinian. As a producer of trade shows, Zimmerman had previously worked with the fair. In 1961, he staged his initial Southeastern Flower and Garden Show at the fairgrounds and returned the following year. In 1963, however, he moved the show to Atlanta because of limited exhibit space. He began the Southern Farm Show there in 1969, again in large part because North Carolina's state fairgrounds lacked the exhibit space to accommodate such a major

exhibitors from agricultural and related industries and thus attracts primarily an audience directly involved in the state's agricultural economy. More than three hundred exhibitors representing more than five hundred manufacturers display their wares at the show. The displays cover the entire spectrum of the agricultural economy—tractors, farm implements, seeds, chemicals, tobacco equipment and barns, livestock equipment, fencing, irrigation units, hay equipment, and more and resemble the fair exhibits of the middle decades of the twentieth century. The

The Southern Farm Show, held at the state fairgrounds every February, is billed as the "Largest Agricultural Exposition in the Carolinas and Virginia." Pictured is an overview of exhibits in the Jim Graham Building at the 2002 show. Photo courtesy of Zimmerman.

Left: The Southern Farm Show is open to agricultural exhibitors only. These farmers are admiring equipment at the 2002 show. Photo courtesy of Zimmerman.

Below: The state fair's exhibit halls will not contain the huge Southern Farm Show, which spills out onto the fairgrounds. These outdoor exhibits appeared at the 2002 show. Photo courtesy of Zimmerman.

enormous trade show exemplifies the realization of J. S. Dorton's dream of the fair as a year-round exhibition facility.

The manner in which the fair sought to improve farm life underwent change. The fair continued to award farmers its traditional premiums for the best exhibits of a variety of crops, with emphasis remaining on the cash crops of cotton and tobacco. But that emphasis quickly shifted from the quality or quantity of crops raised by individuals to group efforts

to improve farm life. That trend had actually begun during the final years in which the North Carolina State Agricultural Society had administered the fair. At the 1925 fair, for example, six county exhibits carried out the theme of "the improved rural home" by emphasizing products grown in each county. Cleveland County's exhibit demonstrated "the remarkable development made by this county in making rural life more attractive. Over 700 buildings have been recently painted . . .

90 miles of light lines established in 13 communities and 473 homes lighted as a result." The exhibit also highlighted cooperative marketing of farm products, crop rotation, and the correct use of commercial fertilizers.

Established as a division of the Department of Agriculture under the direction of J. S. Dorton in 1937, the fair quickly focused on demonstrating how groups and communities could implement the state's strategy for

economic development and the role of agriculture in those development plans. In 1938 the fair added to its listing of regular exhibit categories the "County Progress" exhibits. The displays were clearly designed to demonstrate how counties might strike a balance between agriculture, industry, and education and to seek to progress in all three areas. The fair's management expressed the hope that "each exhibit entered shall be a miniature

Since North Carolina began to promote agricultural diversity in the 1920s, nursery products have become important components of the state's total farm income. North Carolina's nurserymen enter their plants in a variety of categories at the state fair each year. These flowering plants were displayed at the 1996 fair. Photos courtesy of NCDA&CS.

exposition of the three great industries of each county—pleasing to the eye and attractively arranged." "If there is balanced agriculture in the county," the managers added, "it would be well to prove this. A diversified farming system with gardens, dairying, hog growing, and so on balanced with cash crops and good farming methods is always good." Counties were invited to enter: the first-place exhibit received a $650 premium, the second-place winner, $500, and the third-place recipient, $300—enormous sums in the depression-era economy.

In 1940 management dedicated the fair to the purposes and principles of the Campaign for Balanced Prosperity in the South, 1940-1950, an initiative sponsored by the Southern Governors Conference. The campaign, aimed at breaking the South's reliance upon "money crops," emphasized the need for industrialization and diversified farming. Land-grant colleges such as North Carolina State, as well as other agricultural and educational agencies, supported the movement. The first three of the campaign's "Ten Roads to 'Balanced Prosperity'" dealt directly with agriculture: "balancing money crops with food, feed, and fertility crops; balancing crops with livestock,

and production progress with marketing and transportation opportunities without trade barriers."

When the fair resumed after World War II, it replaced the County Progress exhibits with individual farm displays, which continued to promote diversified farming. Such displays, placed within Department A of the fair's premium list, were required to "represent the individual farm and farm activities of the exhibitor." The purpose of the exhibits was to show "a well balanced farm and home program" that reflected "the entire farm operation" and to display balanced farm production with "not too much of any one thing represented." Exhibits "should prove educational so that practical lessons may be drawn from it by farmers and others with reference to farm and home management as well as production and marketing practices."

In 1952 the fair completely revised its exhibit structure, adopting a new format that, with a few minor changes, continues in use. The format categorized exhibits into five major divisions, each including several departments. Division I included exhibits by educational institutions, including 4-H Clubs, as well as those devoted to field crops,

Since its inauguration in 1853, the state fair has continually awarded premiums to a variety of agricultural products judged the best in their category. These bell peppers received a premium at the 1984 fair. Photo courtesy of NCDA&CS.

Fair crowds admiring prize-winning watermelons at the 1996 fair. Photo courtesy of NCDA&CS.

horticulture, bees and honey, and arts and crafts. Division II encompassed the livestock departments: dairy cattle, beef cattle, all-purpose cattle, horse, swine, and sheep. Exhibits of poultry, rabbits, eggs, and dressed turkeys comprised Division III. Division IV was devoted to home demonstration clubs, as well as culinary, clothing, and home furnishings exhibits. Division V encompassed special features, such as the folk festival, wildlife exhibits, exhibits by state and federal agencies, and a variety of other features that were subsequently added to that category.

Department A of the new Division I, a new exhibit called *North Carolina Accepts The Challenge* replaced the post-World War II Individual Farm Displays. Known as the United Agricultural Program and aimed at North Carolina's farms and rural communities, the Challenge called for (1) increased per capita income, (2) greater security, (3) improved educational opportunities, (4) finer spiritual values, (5) stronger community life, and (6) more dignity and contentment in country living. The North Carolina Board of Farm Organizations (NCBFO), comprised of the state Department of Agriculture, the North Carolina Department of Commerce and Development, the North Carolina Farm Bureau Federation, the Farmers Home Administration, the Production and Marketing Administration, the Division of Vocational Training of the North Carolina Department of Public Instruction, the North Carolina

Rural Electrification Authority, the U.S. Soil Conservation Service, the North Carolina Grange, and both the Agricultural Experiment Station and the Agricultural Extension Service of North Carolina State College, developed

"North Carolina Accepts The Challenge"
THROUGH A UNITED AGRICULTURAL PROGRAM

Department "A"

CHALLENGE

Committee in Charge:

MR. L. Y. BALLENTINE, *N. C. Commissioner of Agriculture*
MR. D. S. WEAVER, *Director, N. C. Agricultural Extension Service*
MISS CATHERINE DENNIS, *State Supervisor, Home Economics Education*

As a feature of the North Carolina State Fair, this department is being devoted to the United Agricultural program through which North Carolina has accepted "THE CHALLENGE" for: (1) Increased Per Capita Income; (2) Greater Security; (3) Improved Educational Opportunities; (4) Finer Spiritual Values; (5) Stronger Community Life; and (6) More Dignity and Contentment in Country Living.

Above: *North Carolina Accepts the Challenge* exhibit from a previous fair as pictured in the 1956 *Premium List.*

Left: Prizes for the best displays of varieties of fruit and vegetable crops began with the initial state fair in 1853, and such exhibits remain popular at the modern fair. These fairgoers are examining prize-winning pumpkins at the 1996 fair. Photo courtesy of NCDA&CS.

and sponsored the challenge. Once again, counties, not individuals, were invited to exhibit. The NCBFO and the Challenge program epitomized the efforts of state, federal, and private agencies to develop coordinated economic development policies. The Challenge also epitomized the fair's role as a state agency that both announced the state's economic development programs to the larger public and attempted to enroll the public in their support.

In 1952 the fair began to feature "theme" exhibits, in which clusters of individual displays within one exhibit area demonstrated to fairgoers the significance of a specific sector of the North Carolina economy. The themes chosen usually emphasized the agricultural economy but occasionally highlighted a particular industry. In 1956, for example, dairying was the chosen theme. The fair initiated a milking parlor in which visitors could "see cows milked by the latest mechanical equipment, following the process of the milk through cooling tanks and pasteurizers[,] and then step up to a window and buy a long, cool drink of the luscious product." Other agricultural "industries" treated by thematic exhibits

included small grains, cotton, tobacco, and poultry. One of the functions of such exhibits was to impress the state's growing urban and suburban population with the continued economic significance of the agricultural economy. Explaining the thirty exhibits on "food meats" at the 1957 fair, the fair's management observed that "Most North Carolinians will be greatly surprised to learn to what extent food meat animals are replacing cotton and tobacco as a prime source of farm income and how important the processing and marketing of meat products have become to the state's entire economy." Foods produced in the state, including dairy products, eggs, poultry, pork and beef products, fruits, and vegetables, received thematic treatment in a forty-eight-booth display in 1966.

During the 1970s, the state's livestock industry received special emphasis. At the 1974 fair, the North Carolina Pork Producers Association mounted a major exhibit on the pork industry. Titled *Hog Heaven*, it displayed "as many aspects of the pork industry as possible," among them the "history of the hog, including how the animal came to America and the uses it has been put to over the years." The

The North Carolina Dairy Industry Committee sponsors a milk bar, featured at the fair since 1956. Photo (1996) courtesy of NCDA&CS.

Left: Samples of Neese's pork products, especially the Greensboro company's liver pudding and sausages, are perennial favorites with generations of fairgoers such as these visitors at the 2002 fair. Photo courtesy of the author.

Below: North Carolina country hams are among the more famous of the "food meats" produced in the state. These hams were displayed by the North Carolina Meat Processors Association at the fair around 1989. Photo courtesy of NCDA&CS.

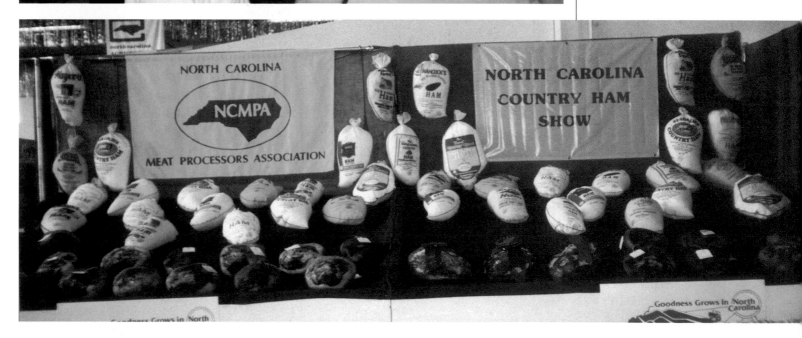

exhibit presaged, and perhaps encouraged, the rapid—some would say too rapid—expansion of the swine industry and its accompanying meat-packing industry in eastern North Carolina during the last two decades of the twentieth century. In 1975 the North Carolina Horse Council presented *The Land of Horses*, an exhibit that employed realistic fiberglass models and photographic displays to trace the history of the horse since its introduction into North America. Again, the exhibit proved a precursor to an even greater emphasis upon the horse industry in North Carolina and at the fair. During the 1980s, industrial themes dominated.

In 2001, however, the fair returned to agricultural concerns with a thematic exhibit titled *Biofrontiers*, designed to show how biotechnology had influenced North Carolina over the two previous decades. The biotechnology exhibit, which ran at the 2001 and 2002 fairs, perfectly illustrated how the fair's function and exhibit philosophy had changed since the mid-twentieth century. While the fair continued to award prizes to individual farmers for the best examples of cotton, corn, tobacco, other field crops, a variety of fruits and vegetables, and numerous household items, such awards were essentially an exercise in nostalgia, a reminder of a time when the subsistence farm had been a reality, not a romanticized ideal. The fair's more pressing agenda was to display the manner in which a vast complex of state, federal, and private

agencies supported the North Carolina agricultural economy and how it in turn was integrated into the state's larger economy, one now dominated by non-agricultural pursuits.

The inauguration of a livestock auction at the 1928 state fair had begun a process of gradual change in the fair's methods of promoting livestock production within the state. The fair continued to award premiums for the best animals exhibited in numerous categories, but such premiums had little impact on encouraging farmers to improve their livestock and poultry. The twentieth-century fair did, however, continue to introduce North Carolina farmers to new breeds and allow farmers to judge for themselves the potential for profits such breeds demonstrated. Such was especially true of cattle exhibits, which in the post-World War II era introduced such new breeds as Brown Swiss dairy cattle and Bahamian, Santa Gertrudis, and Charolais beef cattle to North Carolina farmers. The fair competitively exhibited Charolais cattle for the first time in 1966, for example, with entries coming from North and South Carolina, Virginia, and Georgia.

After the Second World War, the fair increasingly became a livestock market, as well as a means of allowing the state's nonfarm population to see the animals that put meat, dairy products, and eggs on their tables. By the 1950s, the fair held a number of sales of beef cattle and swine, made possible by the construction of Dorton Arena, which had

After the Second World War, the state fair became an increasingly important showcase and market for fine cattle. These champion cows are shown inside Dorton Arena around 1953.
Photo courtesy of NCDA&CS.

been designed with such livestock sales in mind. These shows included sales of calves raised by youngsters in 4-H Clubs and other agricultural organizations. The junior sales, as they were called, became a popular feature of the fair, in part because of their appeal to the nostalgic concepts of farm life, expressed perfectly by Jim Graham in 1980 when he told a *News and Observer* reporter, "I've never known a youngster that raised a cow to go wrong."

Commissioner of Agriculture L. Y. Ballentine established a beef carcass show at the 1964 fair. At the carcass show, "entries are judged on

> *The junior sales, as they were called, became a popular feature of the fair, in part because of their appeal to the nostalgic concepts of farm life, expressed perfectly by Jim Graham in 1980 when he told a reporter, "I've never known a youngster that raised a cow to go wrong."*

Left: These North Carolina youngsters are presenting their cattle for judging at the 2002 fair's Youth Livestock Show. Photo courtesy of the author.

Below: A reserve champion steer at the 1996 fair. Photo courtesy of NCDA&CS.

These cattle are being readied for judging. The 1983 shot (right) is courtesy of NCDA&CS and the 2002 shot (below) is courtesy of the author.

foot. They are then slaughtered and prize winning carcasses will be placed on display in especially refrigerated cases in the cattle barns." Swine were also judged and slaughtered and their carcasses displayed. While the carcass exhibits certainly underscored the relationship between the state's livestock herds and the meat-processing industry, they did so a bit too graphically for the sensibilities of fairgoers, and the exhibits were abandoned in the 1970s.

The fair's emphasis on raising livestock for the market continued, however, with more and more sales events added to the program. The program for 1989, for example, listed a junior steer sale; a junior sales show for steers, market barrows (male swine), and market lambs; a Charolais breeding-cattle show; the "Sale of Champions" (both for cattle and swine); and an all-junior market barrow sale. In addition, the program included a number of shows for specific breeds, among them Seminal, Red Angus, Polled Hereford, Charolais, Santa Gertrudis, Angus, and Hereford beef cattle; Ayshire, Brown Swiss, Holstein, Guernsey, and Jersey dairy cattle; Duroc, Poland China, Spot, Chester White, Tamworth, Berkshire, Hampshire, Yorkshire, and Landrace swine; and for dairy and Nubian goats and sheep, lambs, and ewes. Programs throughout the 1990s duplicated, with very few exceptions, that organizational structure. The 1994 program, for example, added junior ewe and market lamb shows, a Gelbvieh beef cattle show, and a pygmy goat show.

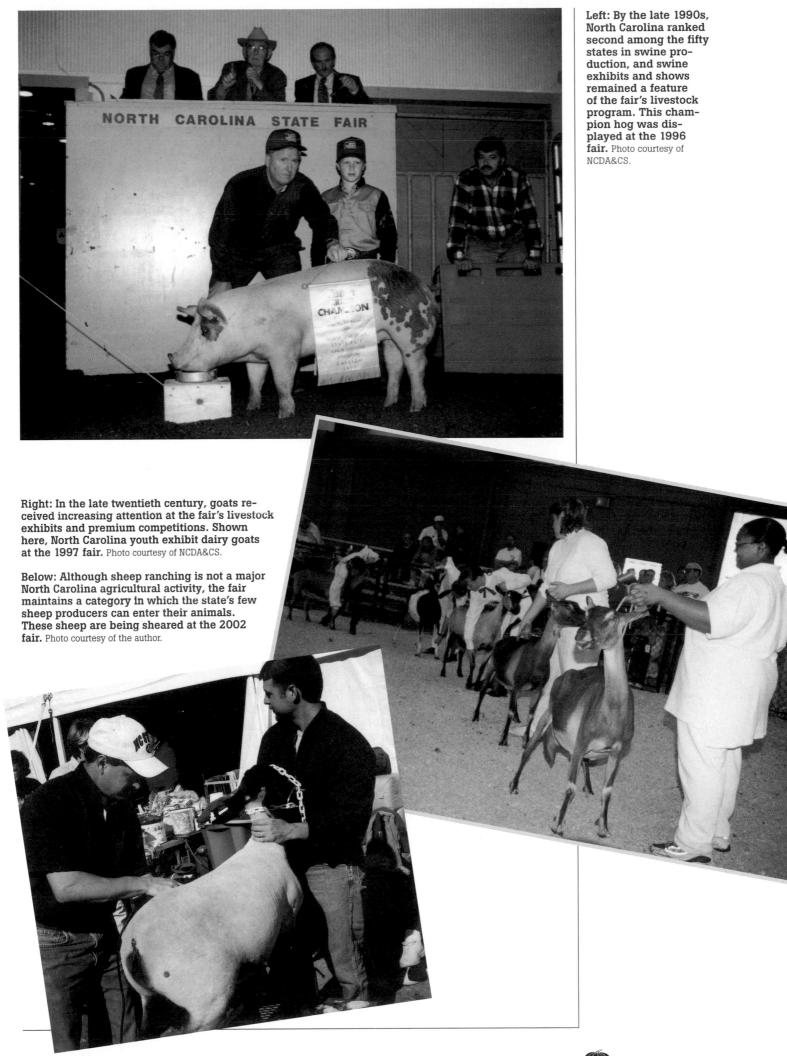

Left: By the late 1990s, North Carolina ranked second among the fifty states in swine production, and swine exhibits and shows remained a feature of the fair's livestock program. This champion hog was displayed at the 1996 fair. Photo courtesy of NCDA&CS.

NORTH CAROLINA STATE FAIR

Right: In the late twentieth century, goats received increasing attention at the fair's livestock exhibits and premium competitions. Shown here, North Carolina youth exhibit dairy goats at the 1997 fair. Photo courtesy of NCDA&CS.

Below: Although sheep ranching is not a major North Carolina agricultural activity, the fair maintains a category in which the state's few sheep producers can enter their animals. These sheep are being sheared at the 2002 fair. Photo courtesy of the author.

Post-World War II prosperity created a thriving horse industry in North Carolina and an increasing emphasis on horse shows at the state fair. This horse show took place in Dorton Arena during the 1982 fair. Photo courtesy of NCDA&CS.

In the post-World War II era, the fair's horse exhibits reflected major changes in the agricultural economies of the state and nation. Once under state control, fair management had emphasized draft animals, rather than the pleasure horses and racehorses favored by the State Agricultural Society. That emphasis continued after the war, but the tractor rapidly replaced horses and mules on North Carolina farms by the end of the 1950s. Meanwhile, the North Carolina horse industry was developing. In 1946 fair manager J. S. Dorton announced the formation of the North Carolina Saddle Horse Breeders Association, an outgrowth of meetings involving Dorton, Commissioner of Agriculture W. Kerr Scott, and North Carolina horse fanciers held in December 1945. The purpose of the organization, which Dorton expected to achieve a full enrollment of three thousand members by April, was to improve saddle horse breeds in North Carolina and to encourage further study of soils and feeds as a means of developing high-caliber horses.

Fueled by the postwar prosperity enjoyed by North Carolinians, who flocked to the state's rapidly expanding urban centers, the North Carolina horse industry took off. Dentists and doctors, lawyers and professors, middle managers and state officials and their children began to spend millions of dollars keeping horses in stables plain and fancy for a weekend of pleasure riding. Breeders began to spend additional millions to produce the animals; agronomists in the state Department of Agriculture and at North Carolina State College began to study how to improve feeds and turf for the state's growing herds, and small businessmen rushed to provide weekend riders with riding tack, saddles, riding costumes, and other paraphernalia.

The fair took note of this burgeoning industry and in 1967, in a move that would have made Edith Vanderbilt proud, reintroduced the horse show. An infinitely more prosperous North Carolina was ready for it. The 1967 show, approved by the American Horse Show Association, exhibited Morgans, Arabians, hunters, American quarter horses, Appaloosas, Tennessee walking horses, American paints, and Shetland and Welsh ponies. It was an instant success, and the fair immediately

Agriculture Jim Graham in 1981 requested from the state legislature more than $4 million to build a modern horse complex at the fairgrounds. Graham's appeal contained nothing

Graham's appeal contained nothing of the nostalgia for the family farm. Rather, it was a hardheaded business proposition. The North Carolina horse industry, Graham informed the legislature, was big business, worth at least $200 million to the state's economy, a figure that tripled to $600 million when the motels, restaurants, gasoline stations, and merchants that catered to horse-show folks were added to the mix.

began to plan for bigger and better shows. By 1979 the fair's annual horse show had become one of the largest in the South, attracting 1,500 horses owned by exhibitors from twenty-three states and awarding $25,000 in premiums.

By 1981 the horse industry had developed into a significant sector of the North Carolina agricultural economy and, unlike others, one that had a large urban and suburban constituency. Convinced that the state fair could become a year-round center for the showing and marketing of horses and the various businesses of the horse industry, Commissioner of

of the nostalgia for the family farm. Rather, it was a hardheaded business proposition. The North Carolina horse industry, Graham informed the legislature, was big business, worth at least $200 million to the state's economy, a figure that tripled to $600 million when the motels, restaurants, gasoline stations, and merchants that catered to horse-show folks were added to the mix. Graham explained that the fourteen shows booked for the fairgrounds in 1980, staged in Dorton Arena and the Graham livestock building, were costly to hold, requiring constant assembly and disassembly

Horses and riders prepare for competition in the Gov. Hunt Horse Complex at the 1996 state fair. Photo courtesy of NCDA&CS.

A lone rider readies her horse for competition outside the Gov. Hunt Horse Complex at the 1996 fair. Photo courtesy of NCDA&CS.

facility with permanent seating for approximately five thousand people, an outdoor warm-up ring, five permanent barns with stalls for more than 900 horses, and temporary stalls for an additional 500 steeds.

Jim Graham's predictions about the horse facility proved inaccurate only in that they underestimated the use it would receive. Since its construction, the facility has become one of the busiest on the fairgrounds, has its own director, and is in constant use year-round. The highlight of the year is, of course, the North Carolina State Fair Horse Show, held each October during fair week. The 2002 fair included events every day, with five days devoted to hunters and jumpers the week before the fair. The 2002 fair staged separate shows for the following horse breeds: Haflingers, paints, Palominos, quarter horses, Appaloosas, walking horses, Arabians and half-Arabians, carriage drivers, roadsters, saddlebreds, Morgans, miniatures, and Paso Finos. Draft animals received but two slots on the schedule—one reserved for Haflingers, a breed of Austrian draft horses unknown to North Carolina farmers before World War II, and one for the lowly but still venerated mule. In the remaining eleven months of 2002, the fair held forty-eight additional horse shows, with at least two scheduled for every month of the year and a half-dozen sched-

of temporary stalls. The new horse show facility at the fairgrounds, Graham assured the legislature, would allow the fair to operate shows efficiently throughout the year and bring additional moneys to the state and especially the Raleigh area. It would also make the fair a profit of at least $13 million in its first year of operation and approximately $100 million by the fourth year of operation, Graham promised. Graham's proposal, endorsed by Gov. James B. Hunt Jr., the secretary of the North Carolina Department of Commerce, and the Raleigh Chamber of Commerce, sailed through the legislature. At the 1983 fair Governor Hunt dedicated the horse complex (named in his honor), an 81,000-square-foot

uled in March, May, and September, barely giving horse-complex crews time to strike one show and set up for the next. The state fair had become the horse-show capital of the Southeast.

The rise, fall, and rebirth of the fair-sponsored horse show in the twentieth century effectively illustrates the changes to North Carolina's economy that have occurred over time, as well as efforts on the part of the state fair to accommodate those changes. The modern horse show, begun in 1921 under the leadership of the Agricultural Society and Edith Vanderbilt, was simply out of place in a North Carolina still dominated by small farmers, many of them tenants, struggling to

Competitions for hitched teams of horses reflected competitors' nostalgia for the rural past. These two teams showed at the 1999 state fair.
Photo courtesy of NCDA&CS.

wrest a living from the land. The economic prosperity that followed the Second World War transformed the state's agriculture, brought rapid mechanization to the farm, hastened the demise of farm tenancy and the small family farm, encouraged the rise of the corporate farm and agribusiness, and witnessed a huge migration of rural people to burgeoning urban areas. By the mid-1960s, a prosperous, largely nonfarm North Carolina population was ready for a horse show. By the 1980s, North Carolinians viewed horses, as they did swine and cattle, tobacco and cotton, as simply just another aspect (albeit a more lovable one) of the commercial-industrial complex that agriculture had become.

A network of interlocking agencies, interest groups, and private industries sustains North Carolina's modern agricultural economy. They include the North Carolina Departments of Agriculture and Commerce, North Carolina State University and North Carolina Agricultural and Technical State University

and their many extension and research programs, the U.S. Department of Agriculture and a panoply of federal farm agencies, agricultural chemical companies, agricultural marketing associations, farm insurance companies, the food-processing industry, the meat-packing industry, and others too numerous to mention. Together, they constitute a formidable bloc, both in the market economy and in the world of politics. In truth, the North Carolina State Fair is no longer necessary to the promotion of agricultural enterprise and in fact can do little to influence it. But it remains a celebration of what rural North Carolinians have accomplished; a showcase of the latest agricultural practices and technology, both for farmers and curious city folk; a connection between North Carolinians on the farm and those in towns and suburbs; a reminder that food and fiber comes from the labor of real people; and, increasingly, a link to the state's agricultural heritage.

N. C.
STATE
EXPOSITION

RALEIGH,
NORTH CAROLINA,
FROM OCT. 1 TO 28, '84

THE FIRST AND ONLY FULL EXHIBIT OF ALL THE STATE'S RESOURCES EVER MADE.

A SPLENDID BAND
Will be in attendance during the month.

GRAND PRIZE DRILL

BAND TOURNAMENT

STATE ENCAMPMENT GUARD

OCT. 1 & 2

RIFLE SHOOTING

MANUFACTURING MACHINERY
IN MOTION.

A COMPLETE DISPLAY OF
AGRICULTURAL
PRODUCTS
— AND —
MACHINERY!
ORES · MINERALS · TIMBERS
FISHES,
ETC., ETC.
NORTH CAROLINIANS
CAN NOT AFFORD
TO MISS THIS GREAT EVENT

This Grand Exposition will also embrace the
MAGNIFICENT
DISPLAY
OF THE
State Department of Agriculture
AS SHOWN IN BOSTON.
The Payment of Premiums by the
N. C. AGRICULTURAL SOCIETY
ON LIVE STOCK AND OTHER ARTICLES.
ANNUAL EXHIBIT AND AWARD OF PREMIUMS
By the North Carolina
COLORED INDUSTRIAL ASSOCIATION

REDUCED RAILROAD RATES

Main Building is the Largest ever erected in the State.

Remember the Time! Opens Oct. 1—Closes Oct. 28.

W. S. PRIMROSE, President.   H. E. FRIES, Secretary.   L. D. HEARTT, Treasurer.

THE RUSSELL & MORGAN PRINTING COMPANY, CINCINNATI.

The Fair Embraces the New South

Industrial Promotion at the Fair

Promotion of Industry at the Nineteenth-Century Fair

Although it was created to promote both agriculture and industry, the North Carolina State Fair concentrated on agriculture, as would be expected in a state with an overwhelmingly agrarian economy. Nevertheless, even the fairs of the antebellum era attempted to obtain industrial exhibits, with only marginal success. Such exhibits reflected the relative insignificance of industry in the South prior to the Civil War. Practically all industrial products available for exhibition were the handiwork of local craftsmen. Examples of such items were buggies; mattresses; and leather goods such as boots, saddles, bridles, and harnesses. Several cabinetmakers displayed their work, many of them Raleigh concerns, although shops from as far away as Salem entered exhibits, as did coach and buggy makers from towns throughout the state.

Exhibits of truly industrial, machine-made North Carolina products were limited to textiles, the state's only developed enterprise during the antebellum era and an industry in which many of the fair's early leaders, such as Thomas Holt, were involved. Seven cotton mills from

---

This poster for the 1884 State Exposition announced to the world North Carolina's intention of complementing its agricultural economy with an industrial capacity. Photo courtesy of A&H.

After the Civil War, tobacco manufacturing challenged textiles as North Carolina's leading industry. This Vance County exhibit at the 1884 State Exposition showcases the products of Joseph E. Pogue's tobacco factory in Henderson. Pogue served for many years as secretary of the North Carolina State Agricultural Society.
Photo courtesy of A&H.

various towns in the state displayed their products at the initial fair, but only one or two textile firms exhibited at subsequent antebellum fairs.

Northern firms dominated the exhibits of machine-made goods. While such exhibits featured textile and leather products, manufacturers of organs and pianos mounted the largest exhibits, hoping to sell their products to fair visitors. North Carolinians noted, and some resented, the dominance of northern firms, which served as a tangible reminder of the state's and the South's dependence upon industrial goods manufactured exclusively in the North.

When the fair was reestablished after the Civil War, the type and number of industrial exhibits displayed there changed little. The fair continued to be almost exclusively an agricultural exposition, and the few industrial items to be seen resembled the crude, handmade goods that characterized the antebellum fairs. Within a few years, however, as leading southern politicians and newspaper editors called for an industrialized South that could compete on equal terms with northern industries and North Carolina's economy underwent changes, the fair began paying increased attention to the subject of industrial development.

During the antebellum period, several small tobacco factories in North Carolina had exhibited their products—mostly plug tobacco but also cigars and smoking tobacco—at the fair. The Civil War introduced Northern soldiers, as well as Confederates from regions of the South outside North Carolina, to bright-leaf tobacco, creating a demand for the state's tobacco products, which in turn resulted in the rapid development of tobacco-manufacturing companies. In the early 1870s several companies began to use the fair to promote their products. The majority of the state's 126 widely scattered tobacco companies never exhibited at the fair, however.

By the late 1880s and early 1890s, large tobacco companies dominated the industry, forcing many of the smaller firms out of business and depriving the fair of its principal exhibitors. For several years the firms of Joseph E. Pogue and W. T. Blackwell, two of the state's largest companies, completely monopolized the tobacco exhibits at the fair. Both Pogue and Julian S. Carr, one of the major owners of the Blackwell firm, also owned large farms and were longtime members of and officers in the North Carolina State Agricultural Society, a circumstance that accounts for their firms' loyalty to the fair. Because of their huge size and national scope, the giants of the industry bypassed the fair as a means of promoting their products. The largest of them, the American Tobacco Company, famous for its extensive national advertising, consistently failed to employ the fair to reach potential consumers in its home state. The company's leaders, members of the Duke family, unlike Julian Carr, an industrialist with a planter background, lacked any affiliation with the Agricultural Society and represented a new element among North Carolina's elite—those who came from humble backgrounds to make fortunes in industry.

Despite the fact that cotton mills, well established prior to the Civil War, were the state's oldest industry, North Carolina textile firms for the most part chose not to exhibit at the fair. In 1840 there were 25 cotton mills in the state, and by 1860 there were 39, but only 7 of them exhibited at the initial 1853 fair. Textile exhibits at the fairs of the antebellum years and the 1870s were limited to different types of yarns, usually of inferior quality. Under the leadership of Thomas M. Holt, himself a cotton textile manufacturer, the State Agricultural Society sought to attract additional textile exhibits to the fair during the 1870s. The cotton textile industry responded modestly to that initiative, and several firms displayed goods much superior to the antebellum exhibits, including various types of

Above: Bull Durham, the trademark for the smoking tobacco manufactured by Durham's Blackwell Tobacco Company and the brainchild of Julian Shakespeare Carr, one of the firm's owners, was one of the best-known trademarks of the late nineteenth century. Carr, a leader of the State Agricultural Society, saw to it that his firm's products were displayed at the state fairs. Shown here is the Bull Durham trademark as it appeared in 1896 advertisements, after the Blackwell firm had been purchased by the American Tobacco Company. Photo courtesy of A&H.

Below: Winston in Forsyth County rivaled Durham as a tobacco-manufacturing center. This exhibit of Thompson & Company, cigar manufacturers, appeared at the 1884 State Exposition. Photo courtesy of A&H.

Cotton textile manu-
facturing, the state's
leading industry both
before and after the
Civil War, produced
several exhibitors at
fairs of the late nine-
teenth century, among
them the Worth Manu-
facturing Company,
whose exhibit at the
1884 State Exposition,
pictured here, fea-
tured the latest looms.
Photo courtesy of A&H.

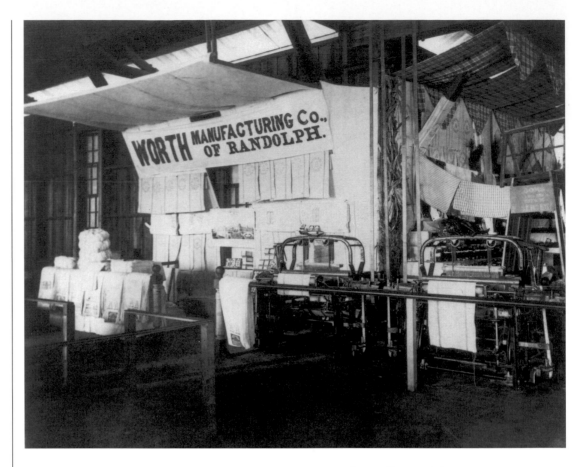

Cotton textile manu-
facturing, the state's
leading industry both
before and after the
Civil War, produced
several exhibitors at
fairs of the late nine-
teenth century, among
them the Worth Manu-
facturing Company,
whose exhibit at the
1884 State Exposition,
pictured here, fea-
tured the latest looms.
Photo courtesy of A&H.

cloth, as well as yarns, carpet warp, twines, and sheeting. Although only a few cotton textile firms exhibited at the fair during the 1890s, the quality of exhibits continued to im-prove as colored cloth, plaids, and finished garments such as socks and underwear were displayed.

Woolen mills, much less numerous, appear to have exhibited at the fair proportionately more than did cotton mills, probably because they manufactured for local markets and saw the fair as an excellent promotional activity. During the 1880s, woolen firms exhibited as many as thirty-four patterns and several types

The state's textile
mills also produced
woolen goods, such as
these displayed by the
F. and H. Fries Com-
pany of Salem at the
1884 State Exposition.
Photo courtesy of A&H.

of woolen cloth. In the 1890s, firms displayed finished products such as blankets and underwear. Major exhibitors of woolen products included the Pothania Woolen Mills, the Mount Airy Woolen Mills, and the woolen mills of the Fries brothers of Salem.

In 1893 the nation entered a severe and prolonged economic depression, resulting in a diminution of textile and other industrial exhibits at the fair. In 1896 Bennehan Cameron, then president of the State Agricultural Society, sought to regain some of the fair's lost industrial exhibitors and to acquire new ones. He approached H. S. Chadwick, president of the Charlotte Machine Company, with a request that Chadwick use his influence with the various textile companies of the state to obtain exhibits for the fair. Chadwick warned Cameron that success would be difficult because of the depression, and his warning proved sound—industrial exhibits did not increase. Indeed, within a month of the fair Cameron was assured of but a single exhibit from a textile firm. Even in prosperous years, the state's textile firms rarely mounted exhibits at the fair. Those that did were owned entirely or in part by members of the Agricultural Society.

The furniture-manufacturing industry was the third major North Carolina enterprise to exhibit its products at the nineteenth-century fair. Beginning with the fairs of the antebellum years, local cabinet shops entered such exhibits and continued to do so after the fair was reinstituted in 1869. Not until the 1890s did larger firms begin to exhibit their products. During that decade, machine-made, factory-produced furniture replaced exhibits of handcrafted items, reflecting the rapid growth and modernization of the state's furniture industry. In 1892 Ruffin Jones of Raleigh exhibited what the press called the first furniture made by machine in the state. The furniture, made of native oak and walnut, was reported to be the equal of any of northern manufacture. Several other firms exhibited during the following years, and in 1899 an exhibit mounted by a group of manufacturers located in High Point—the largest such display at the fair by one industry up to that time—further reflected the growth of the industry. Among the items displayed by the eight factories exhibiting were sideboards, chiffoniers, chamber sets, tables, hall racks, upholstered goods of several types, showcases, trunks, chairs of several grades and types,

The state's fledgling furniture industry relied upon wood produced by steam-powered sawmills that by the end of the nineteenth century dotted the pine forests of the Coastal Plain and the hardwood stands of the Piedmont and Mountain sections of the state. This steam-powered sawmill, restored to running condition for the 2002 fair, was employed in North Carolina forests during the late 1880s and early 1900s. Photo courtesy of Blankinship.

office furniture, beds, and dressers. A higher proportion of North Carolina's furniture factories exhibited products at nineteenth-century state fairs than did firms in the older tobacco or textile industries. Although it expanded rapidly, the furniture industry produced for a more local market than did the other two, and thus viewed the fair as an effective advertising strategy. As with the textiles and tobacco, however, the fair did little to promote the growth of the industry and instead merely reflected that growth. But it did help awaken North Carolinians to the industry and allowed them to take pride in its development.

While the tobacco, cotton, textile, and furniture industry transformed North Carolina's agricultural and forest products into finished goods, the fertilizer-manufacturing industry was directly linked to the production of agricultural commodities. Both native and out-of-state fertilizer firms exhibited at the post-Civil War fairs. The Navassa Guano Company of Wilmington, which imported the South American product, exhibited at the fair in 1870, as did a fish guano firm from Norfolk.

The Navassa firm exhibited periodically for the remainder of the century, awarding premiums for the best crops grown with its products, a practice soon adopted by several other fertilizer firms and dealers. Those exhibits helped convince North Carolina farmers of the value of commercial fertilizers, contributing both to company profits and improved agricultural practices.

Fertilizer exhibits also helped awaken North Carolinians to the potential of commercial fishing to provide a resource for the manufacture of fertilizer. In 1881 the Upshur Guano Company exhibited a specimen of fish scrap from Beaufort that the company agent described as the best purchased by his company that year. Among additional fertilizer firms exhibiting at the fair during the 1880s and 1890s were the Durham Fertilizer Company, the Canton Chemical Company, the Raleigh Oil Mill and Fertilizer Company, the Caraleigh Phosphate Mills, and the Virginia-Carolina Chemical Company. As in the tobacco and textiles industries, a personal link often existed between the fair and exhibiting firms. William Upchurch, a leader in the State Agricultural Society and at one time its president, was likewise president of the Raleigh Oil Mill and Fertilizer Company, one of the fair's largest exhibitors. A prophetic reporter noted that the fish-scrap business "promises to be a great industry in the future of North Carolina." The increased use of fishmeal in commercial fertilizers led to the creation of a new industry along the state's coast. Throughout the first half of the twentieth century, hundreds of boats from ports such as Beaufort, Swansboro, and Southport set out to capture menhaden where they were transformed into fertilizer in on-shore factories.

A variety of smaller industries continued to exhibit at the post-Civil War fairs. Carriage and buggy manufacturers from several towns were usually well represented. The Tyson and Jones firm of Carthage, the Piedmont Wagon Company of Hickory, and the Hackney Brothers of Wilson were among the largest exhibitors of carriages. Firms from several towns exhibited leather goods, but the work of Elias F. Wyatt and Son dominated the competition in that field from the 1880s until the close of the century. The food-processing industry exhibited at fairs during the 1880s, the largest exhibits coming from Forsyth County.

By the late nineteenth century, North Carolina farmers were using large quantities of imported guano to fertilize their fields. A perennial exhibitor at the late-nineteenth- and early-twentieth-century fairs was the Navassa Guano Company, located outside Wilmington. This exhibit appeared at the 1884 State Exposition. Photo courtesy of A&H.

From the 1880s until the end of the century, Edwards, Broughton & Company of Raleigh usually represented the printing industry. Additional industries represented at the fair at one time or other during the post-Civil War period included quarrying, sawmilling, pottery making, shoe manufacturing, tile and brick manufacturing, and the manufacturing of grain and fertilizer bags. Local firms exhibited wood products such as caskets, coffins, baskets, and building materials. Among the most unusual native products exhibited at the fair were pine-straw carpets made by the Acme Manufacturing Company of Wilmington, suggesting an almost desperate search by North Carolina entrepreneurs for a profitable industrial product.

Above left: Several buggy manufacturers and retailers exhibited at the state fair well into the twentieth century. This advertisement for buggies was reproduced from the 1916 *Premium List*.

Above right: The Raleigh printing firm of Edwards, Broughton & Company was a loyal exhibitor at late-nineteenth- and early-twentieth-century state fairs, at which they displayed their latest printing presses and processes. This exhibit appeared at the 1884 State Exposition. Photo courtesy of A&H.

Left: Pine products remained a major North Carolina industry in the late nineteenth century. The state Department of Agriculture demonstrated the use of a turpentine still at the 1884 State Exposition. Photo courtesy of A&H.

The Whitin Machine Works displayed its products at the 1884 State Exposition. Photo courtesy of A&H.

After the mid-1870s, the fair became a commercial center, with dealers hawking a variety of merchandise to fairgoers from every corner of the state. Merchants who held dealerships for products manufactured outside the state were particularly active, displaying stoves, shoes, hardware items, guns, dry goods, sewing machines, and other articles. Dealers in musical instruments, especially organs and pianos, tried to convince fairgoers to bring music and culture into their homes. Agents from the Fleischmann Company of New York exhibited their yeast, and the Kingam Company of Richmond displayed their hams, meats, and lard.

Since the fair's leaders were more interested in agricultural than in industrial matters, external factors largely influenced the fair's program to promote industry. In the antebellum period, industrial exhibits had been meager because little industry existed in the state. The election in 1873 of Thomas M. Holt, himself a manufacturer, as president of the Agricultural Society reflected the idea of the new industrial South that had captivated the minds of many southerners. Holt placed greater emphasis on industrial promotion into the 1880s, and the number of industrial exhibits increased until the 1890s, when a nationwide depression curtailed efforts to promote industry and led to

Bennehan Cameron's increased efforts in 1896 to attract exhibits from the factories of both North Carolina and other states. As better times returned, the number of industrial exhibits at the fair again increased, and by the turn of the century the fair showed signs of becoming a significant industrial exhibition.

The society's charter assigned the fair the task of promoting industry when industry, like agriculture, was an individual concern. The state's antebellum factories, especially in the textile and tobacco industries, were largely established and personally directed by successful planters who recognized the advantage of finished products over raw crops. The fair could never hope to promote industrial development, as it could agricultural production, by encouraging the production of certain types of goods, the adoption of new methods of production, or the use of new machinery. It lacked the means to accomplish so huge a task. Premiums offered for crop production appealed to the individual farmer; premiums offered for industrial production meant nothing to the industrialists. The fair could test most farm machinery; it was impossible to test the heavy machinery required by industry. Farming in the nineteenth century was an individual, small-scale enterprise requiring little capital, and the fair could instruct the

farmer in the use of the best agricultural techniques and equipment. The fair could do little to promote large-scale industry, but it could and did encourage the development of small-scale industry and handicrafts and provided retailers of consumer products an opportunity to introduce their products to North Carolinians from throughout the state.

Although the majority of the firms in any one industry never participated in the fair, a few firms from each industry in the state exhibited their products. When the tobacco industry began to emerge, several firms displayed their products at the fair. The same was true for the textile, furniture, and fertilizer industries. In this manner the fair enabled the people of the state, especially those of the east and west (since most industrial development occurred in the Piedmont and the central section of the state) to realize the importance of the state's budding industries.

In the twentieth century, the state fair continued to serve as a means of showcasing North Carolina's industrial development,

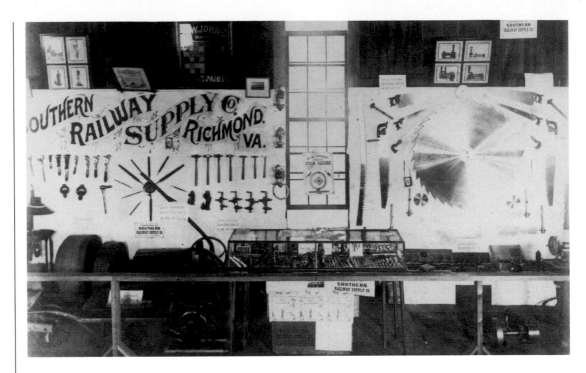

Railroads moved people and products in late-nineteenth-century North Carolina and were among the state's major industries. Pictured here is the Southern Railway Supply Company exhibit at the 1884 State Exposition. Photo courtesy of A&H.

rather than actually promoting that development. The state fair simply lacked the ability to influence the development of modern industrial firms, with their huge and complex factories, their dependence on production machinery far too large to be easily transported for annual exhibition, and their requirement for enormous amounts of capital. Not even by attracting large numbers of workers could the fair have influenced industrial development in the state, for North Carolina's political and industrial leaders fiercely defended laissez-faire capitalism and insisted that factory owners alone, not industrial workers, make all decisions concerning industrial operations. Larger market forces—the availability and costs of capital, the availability of a labor force and the educational levels of that work force, transportation systems, and access to raw materials being the most significant—determined industrial developments in North Carolina's twentieth-century economy. The state fair could and did help acquaint North Carolinians with those developments.

Promotion of Industry at the Twentieth-Century Fair

As in the late nineteenth century, North Carolina State Fairs of the early twentieth century slighted the promotion of industry, instead placing ongoing emphasis on agriculture. The premium list for the 1900 fair continued the practice of devoting but one of more than a dozen exhibit departments, or categories, to manufactures. The program listed departments by alphabetical designation. Manufactures were represented by Department F, which was subdivided into several categories, the most important being vehicles, including buggies, wagons, and drays; furniture, encompassing parlor furniture, beds, and chests; leather goods, including saddles and harnesses; textiles, consisting of the largest and best display by a North Carolina cotton factory; carpentry, including doors, blinds, and mantels; pottery; heating systems; and a miscellaneous category that included wheel rims and spindles. In belated and obviously hastily formed recognition of a then-nascent industrial product, the fair's management first added the automobile to Department F, awarding a gold medal for the best "Automobile or motor carriage" exhibited. With minor exceptions, the fair, while it remained under the control of the State Agricultural Society, continued to exhibit industrial products in Department F. Prior to the First World War, however, the society ceased offering cash awards for industrial exhibits and instead offered diplomas to prize-winning exhibits. In recognition of important changes in the American industrial economy, particularly in the rapid growth of the automobile industry, the fair modified its exhibit categories. In 1915, for example, the fair added "motor delivery trucks" to the exhibit categories in Department F, and by 1920 the category had

FRANKLIN AUTOMOBILES

Another Franklin achievement: a full-size five-passenger Touring Car weighing only 2,280 pounds—$1,850.

Willard Storage Batteries   Rayfield Carburetors
Gray & Davis Electrical Systems

Raleigh Motor Car and Machine Company

L. McA. GOODWIN, Mgr.   (Distributors)   RALEIGH, N. C.

Well before the First World War, automobile dealers were exhibiting their products at the state fair, as this advertisement reproduced from the 1916 *Premium List* illustrates.

grown to include limousines, touring cars, runabouts, electric cars, delivery wagons, auto trucks, auto fire apparatus, motorcycles, and station wagons.

Gradually, the fair's custom of awarding premiums to encourage mechanical invention gave way to the practice of allowing manufacturers or their representatives to display to fairgoers the products of a modern industrial economy, most of which were manufactured outside the state or the South. Again, the automobile serves as an example of that process. Transformed in the first decade of

the twentieth century from the handcrafted product of local buggy shops into the ultimate product of the modern factory assembly line, the automobile epitomized the emergence of a manufacturing economy so vast and complex as to be completely and irrevocably beyond the influence of the fair. The fair audience, on the other hand, attracted auto dealers anxious to display and sell their products. At the 1910 fair, for example, the Raleigh Motor Car and Machine Company exhibited the Jackson automobile, and the United Motor Company of Charlotte exhibited both the Maxwell and

This 1916 photo suggests that automobiles were filling the fair's parking lots, even at that early date.
Reproduced from the 1916 *Premium List.*

Chalmers automobiles, which, the company declared, were "cheaper to operate than a horse and buggy."

In 1920 the Agricultural Society formally abandoned the notion that its awards could influence industrial development. Noting the "impracticality of establishing a system of judging automobiles, machinery, mechanical and manufactured exhibits on their merits," the fair's management officially adopted a policy of providing display space for retailers of industrial products for a "nominal rental." The fair continued to award diplomas, however, to a wide range of manufactured items exhibited, including textiles and leather goods made in North Carolina. The textile category was expanded to include cassimere or jeans and hosiery, reflecting the early beginnings of an apparel industry within the state. Additional products on exhibit included automobiles and other vehicles, cabinetry, sewing machines, typewriters, furniture, china, jewelry, toys, hats and caps, carpets and rugs, books, and ironwork.

At the 1925 fair, the last held under the auspices of the Agricultural Society, major industrial exhibitors included Raleigh's Saunders Motors, which displayed Ford, Fordson, and Lincoln automobiles, as well as Ford trucks, and International Harvester Company, which showcased its latest school bus. The J. C. Benjamin Company and Henderson's Corbitt Motor Truck Company displayed heavy road-building equipment, including wheel graders and heavy trucks, and the DeWalt Products Company of Pennsylvania exhibited its line of woodworking equipment. The American Saw Mill Machinery Company displayed nine

BROADCASTING DIRECT FROM THE STATE FAIR BY WPTF

BE SURE TO CALL AT OUR STUDIO
In the Main Exhibition Hall

Radio advertising is the new medium for the up-to-date business man—Reasonable in price, yet effective.

WPTF   **Raleigh, N. C.**

machines, and the Frick Company exhibited a portable sawmill and a variety of steam engines. The infant electrical appliance industry was represented by Raleigh Electric Service Company's line of Delco and Frigidaire products, and Winston-Salem's Clinard Electric Company exhibited lamps and refrigerators that "made ice." At the dawning of the electric age, when most rural areas lacked electric service, both firms also exhibited "light plants," designed to generate electricity for individual homes. Other firms displayed metal culverts, pumps, adding machines, soft-drink dispensers, radios, a variety of electrical appliances, rugs and carpets, and a host of other domestic products. By the mid-1920s, the fair had entered the age of the consumer and had become a venue through which industrial America could tempt fairgoers with its latest device designed to make work less difficult and life more enjoyable.

When the state Department of Agriculture assumed control of the state fair in 1928, it quickly made clear that its primary goal for the fair was agricultural promotion and that it had no interest in promoting the development of specific industries. The 1928 fair relied upon the Agricultural Society's premium list (including Department F), but in 1929 it dropped manufactures as a category. From that point forward, the fair remained a means for displaying the state's industrial progress through the commercial exhibits of a variety of retailers.

large measure thanks to the success of the New Deal's Rural Electrification Administration, which brought inexpensive, reliable electric power to practically all of rural North Carolina. Retail exhibitors demonstrated refrigerators, home freezers, washing machines, dryers, vacuum cleaners, and electric ranges to fairgoers anxious to adopt a modern lifestyle, and electric utilities such as Carolina Power and Light (CP&L) made Reddy Kilowatt®, a cartoon character symbol of the industry, a familiar figure. In 1967, for example, CP&L, Duke Power Company, and Virginia

Fairgoers browse at commercial exhibits that line the aisles of the Gov. W. Kerr Scott Building, the fair's largest commercial exhibit hall, during the 1983 fair. Photo courtesy of NCDA&CS.

When the state Department of Agriculture assumed control of the state fair in 1928, it quickly made clear that its primary goal for the fair was agricultural promotion and that it had no interest in promoting the development of specific industries.

Especially after World War II, as Americans and North Carolinians began a two-decade spending binge fueled by pent-up demand, an industrial plant unrivaled in the world, and unprecedented prosperity, vendors of consumer products flocked to the fair. Big-ticket electrical appliances took center stage, in

Electric and Power Company operated an all-electric kitchen at the fair's Food Festival. Other exhibitors displayed products ranging from encyclopedias to Fuller brushes to the hundreds of thousands who roamed the corridors of the fair's growing number of exhibit halls.

Commercial exhibits followed the pattern established by agricultural implements and automobiles, and for the same reason. Gradually, as improved transportation and distribution systems placed retailers of appliances and other domestic items within an easy drive of even the most remote rural areas, the fair lost its appeal as a market for such exhibitors. As the traditional displays of electrical appliances and other household items declined, exhibits of products not universally available or just being introduced to the market increased, and a number of what can be described as "luxury" items began to be featured. In 1958, for example, the fair staged its first boat show, which featured more than one hundred boat models. The show was staged not only to provide North Carolinians the opportunity of purchasing a boat but also to alert them to the fact that boatbuilding had become a major industry within the state. At recent fairs, swimming pools and hot tubs have occupied considerable exhibit space, providing further evidence of the state's economic prosperity. In general, however, the fair is no longer a major venue for the promotion of consumer products.

While the fair abolished its "Manufactures" exhibit category in 1928, it did not abandon its commitment to industrial development. The encouragement of industrialization had become a major theme of state economic development policy, and the fair (by then a state agency) became an even more forceful advocate of that initiative. The 1938 fair inaugurated the "County Progress" exhibits, designed to encourage North Carolinians to strive for an economy balanced between agriculture, industry, and education. Employing those exhibits, which continued until 1941, fair management invited up to ten counties to mount exhibits. Exhibitors were specifically instructed to display their county's progress in attracting major industries and commercial development.

In 1940 the fair adopted as its theme "For Balanced Prosperity in the South, 1940–1950," a program sponsored by the Southern Governors Conference and cooperating committees of citizens and public agencies. Begun in January of 1940, the campaign enrolled the press, leading politicians, and land grant colleges, and stressed that the region must industrialize and called specifically for farms to be balanced by factories. When the fair adopted its "North Carolina Accepts the Challenge" program as an exhibit category in 1952, the need for industrialization remained a major theme. Continued until 1969, Challenge exhibits were mounted by counties invited to do so by the

Much of the fair's present promotion of large consumer items, such as boats, camping equipment, or recreational vehicles, is done not during fair week but at trade shows scheduled throughout the year in the fair's major exhibit halls. Typical is this boat show held in Dorton Arena in 1982.
Photo courtesy of NCDA&CS.

fair's management. In 1969 "Community Programs," sponsored by the North Carolina Board of Farm Organizations and Agencies, replaced the Challenge exhibits, but the concept remained basically the same. Community Programs exhibits, staged by invited communities, were more focused on agriculture, but industrial developments remained important. The displays were designed "to illustrate certain phases of their community programs as a means of inspiring and motivating other communities to take action." Community Programs exhibits remained a feature of the fair for a quarter-century, making their final appearance in 1994.

In 1952 management began the practice of announcing a central theme for each fair. While agricultural themes dominated, some fairs were devoted to industrial themes and others to themes that embraced both agriculture and industry. In 1955, for example, cotton was the theme, and exhibits addressed that product's importance to the state's industrial and agricultural economies. Lois Faulkner, the National Maid of Cotton, presented her "fabulous wardrobe of summer and winter cottons" at least seven times a day to fair audiences, a theme revisited in a 1974 exhibit, which emphasized new uses for cotton fibers through fashion shows. In 1963 forestry and wood products received the thematic treatment. In his speech to opening-day crowds, Gov. Terry Sanford informed fairgoers that North Carolina ranked first in the nation in the manufacture of wooden furniture, hardwood veneers,

and hardwood plywood and noted that in forty-one of North Carolina's one hundred counties, more than 60 percent of the land area was covered by trees. The 1972 fair featured *Returns from the Future*, an exhibit that informed fairgoers of advances in space exploration and included a model of the Apollo 12 command module, astronaut suits, and other National Aeronautics and Space Administration items. The 1980 fair featured an exhibit by eight North Carolina governmental agencies, including the Department of Labor and the Employment Security Commission, that demonstrated "the cooperative effort of the agencies involved to give people jobs and

People viewing DNA extraction at *Biofrontiers* (2001). Photo courtesy of NCDA&CS.

Above and below: Since the late nineteenth century, fair exhibits have portrayed educational institutions, especially colleges, as promoters of industrial development, or at least as contributors to it. These exhibits, mounted for the State Exposition of 1884 by Peace Institute of Raleigh and Salem Academy of Salem (now Winston-Salem), reflect an increased emphasis on the education of women. Photos courtesy of A&H.

to make sure working conditions are safe." From 1995 to 2000, the fair hosted *Cyberspace*, an exhibit devoted to showcasing the World Wide Web and recent advances in computer technology to fairgoers. Fittingly, the fair began the twenty-first century with a two-year exhibit called *Biofrontiers* about the state's rapidly developing biotechnology industry.

Education, while not an industry, was also promoted by the fair, under both the Agricultural Society and the Department of Agriculture. Nineteenth-century fairs awarded a variety of premiums for educational exhibits. The early-twentieth-century fairs of the Agricultural Society had a department for educational exhibits, and in 1915, for example, a number of public schools exhibited at the fair, including Waynesville High School and the North Carolina School for the Blind. When the society reorganized its premium schedule in 1924, it dropped the general educational category and, in keeping with the tenor of the times, replaced it with categories for Vocational Agricultural Schools and Boys and Girls Clubs. The County Progress exhibits expressly called for displays to address new and progressive ideas in education in both grammar and high schools. For the 1937 fair, manager J. S. Dorton restored a special educational exhibit category that was specifically designed to portray "some of the phases of the activities of children in the elementary and high schools of the State in other than vocational fields."

In 1952, when the fair reorganized its exhibit schedule, all educational categories were placed in Division I, General Exhibits. Four of the categories dealt directly with education. Category A of Division I, "North Carolina Accepts the Challenge," listed improved educational opportunities as one of the criteria to be addressed by exhibits entered. Challenge exhibits with their emphasis on education continued until 1969. The fair also placed several other educational categories in Division I, including Vocational Education and, for the first time, a category reserved specifically for the North Carolina Department of Public Instruction. In 1954 the fair announced that "the state's new educational television system, WUNC-TV," would make its first-ever telecast from Dorton Arena, thus introducing both fairgoers and the home audience to what would quickly become a North Carolina educational institution.

Education was the theme of the 1962 fair, reflecting the emphasis of the political program of then governor Terry Sanford, and opening-day ceremonies saluted all levels of education in the state. Exhibits on the public schools highlighted new and old methods of instruction and the rising cost of education, and, in a shameless display of partisan politics, the fair staged special tributes to former governor Charles B. Aycock, the "educational governor" of the turn of the twentieth century, and to Governor Sanford.

By the early 1970s, the state's community college system had joined the Department of Public Instruction in the special-exhibit category. Additional displays by a variety of other educational agencies, both public and private, continue to remind North Carolinians of the necessity of continued educational progress. Going into the twenty-first century, the fair remains a crucial agency for acquainting North Carolinians with the manner in which education fits into the state's overall plans for economic development.

The modern fair's efforts at promoting industry are drastically changed from those of the generation of leaders who founded the fair in 1853. They are even further removed from those of the fair's leadership of the New South era. Leaders such as Thomas Holt, Bennehan Cameron, Julian S. Carr, and Joseph Pogue hoped that the fair could stimulate the state's industrial development and help lead it out of an economic wilderness. Given the nature of complex industrial societies, it was an impossible dream. But the fair could, and did, interpret for North Carolinians the enormous economic forces that transformed the state's economy in the middle years of the twentieth century. It could, and did, encourage them to support efforts to make industrialization a priority, just as it displayed to them the benefits of life in an industrial society. It could, and did, sponsor exhibits that encouraged North Carolinians to support educational advances, exhibits that underscored the connection between education and economic progress. As a state agency for most of the twentieth century, the North Carolina State Fair proved a valuable means of interpreting the state's economic development policies to a large segment of the state's citizenry.

Conclusion

It is almost impossible to summarize the impact of the North Carolina State Fair on the lives of North Carolinians since it began in 1853. The sheer numbers of fairgoers is staggering. During the nineteenth century, when the average fair ran for only five days each year and had an average daily attendance of more than 6,000 persons, at least 1.11 million persons attended the thirty-seven fairs held before 1900. Measured against the state's 1900 population of 1,893,810, that figure is all the more impressive. Attendance figures climbed steadily during the twentieth century and reached truly astounding numbers in the post-World War II era. For example, in the years from 1988 to 2001, during which time the fair ran for ten days each year, attendance averaged 707,522 persons per year, meaning that 9,905,308 people attended the fair in that fourteen-year period alone. While it is almost impossible to determine how many individual Tar Heels have attended the fair, inasmuch as some attended every year and others have attended once in a lifetime, based on the attendance figures of the last fourteen years it seems reasonable to assume that as many as two million individuals attended the fair in that period, or approximately one-quarter of the state's current population.

For generations, fireworks exploding over the grandstand infield have signaled the close of exhibits at the fair. For many families this dazzling display means that it is time to head home. Pictured here are fireworks from the 1996 fair. Photo courtesy of NCDA&CS.

These crowds are lined up outside the fairgrounds' gates to purchase tickets for the 1970 fair. Photo courtesy of *N&O*.

Those attending the fair came from all sections of the state, from rural regions and urban and suburban communities, and from all segments of the state's social structure. For a century and a half, literally millions of fairgoers have been exposed to exhibits of livestock, field crops, farm implements, consumer products, industrial products, and efforts to advance community development. To assume that these exhibits had no influence upon them would be sheer folly. Nevertheless, it is extremely difficult to determine to what

Attendance figures do, however, reveal one indisputable fact: North Carolinians have enjoyed, and continue to enjoy, going to the fair.

The fair's influence on the state's economic development throughout the nineteenth century is more easily demonstrated. The history of the nineteenth-century fair, in terms of its economic influence, can be divided roughly into four distinct periods. Each period coincides, to a large degree, with a corresponding stage in the state's general agricultural development. The fair both encouraged and

For a century and a half, literally millions of fairgoers have been exposed to exhibits of livestock, field crops, farm implements, consumer products, industrial products, and efforts to advance community development. To assume that these exhibits had no influence upon them would be sheer folly.

extent the fair influenced the economic decisions of those who attended, and even more difficult to determine its impact upon the state's economic growth and development.

reflected that general development, exerting more influence on some of its particular aspects than upon others. Within each of those stages the fair made some attempt to promote the state's industrial growth, but those attempts were of markedly secondary nature, as agricultural promotion was always its primary goal. The four stages into which the nineteenth-century fair can be divided are as follows: the antebellum period (1853–1860), which coincides with the agricultural renaissance that occurred within the state during the 1850s; the immediate postwar period (1869–1873), which coincides roughly with the period of agricultural reconstruction following the Civil War; the period of Thomas M. Holt's presidency (1873–1884), which corresponds with the years in which the foundations of modern, commercial, and scientific farming were laid; and the post-Holt years (1885–1899), which correspond with a period of continued improvements in agricultural methods coupled with farmers' increased efforts to improve the state's agricultural conditions through political action.

From its inauguration in the decade before the Civil War into the 1890s, the state fair played a significant role in North Carolina's agricultural development. With the exception of a few agricultural journals, the fair represented the only statewide medium through which the farmer could become acquainted with and be instructed in methods of agricultural reform. Premiums for essays

and experiments pertaining to agricultural subjects, annual addresses, and lectures at the North Carolina State Agricultural Society's nightly meetings instructed the farmer in the use of scientific agricultural methods. Such attempts at instruction proved too stiff and formal to appeal to the average farmer and were limited in their effectiveness. They did, however, encourage the exchange of ideas between the state's more substantial and progressive farmers and planters.

The fair's major contribution to the state's nineteenth-century economy was the interest in agricultural reform that it engendered. Exhibits of livestock, field crops, vegetables, farm implements, and other items created that interest. This was especially true of the exhibits of cattle, swine, and horses. Viewing those exhibits forced farmers to make a mental comparison of the stock they saw displayed with the stock on their farms, generating at least an interest in improved livestock.

The crop-lien system, which forced farmers and planters to sow and raise a money crop, increased tremendously the production of cotton and tobacco. The state fair reinforced that system by recommending the use of improved seeds, better cultivating practices, and commercial fertilizers in the production of those crops and awarding large premiums for

Chairs on a Skyway provided an excellent overview of the fair around 1970. Photo courtesy of *N&O*.

the best per-acre yields of each. Exhibits of farm implements by both native and out-of-state manufacturers and dealers acquainted farmers with the latest in agricultural machinery. The fair exhibited implements from several states, especially from the states of the northern Midwest, and placed special emphasis on the development of new and improved implements. The fair also tested all implements displayed for the benefit of the farmer. By the end of the twentieth century the number of implements exhibited had increased substantially, and several big-name companies such as John Deere and McCormick had begun to enter exhibits. Through its exhibits, the fair served as a medium through which thousands of farmers from all sections of the state became acquainted with the latest and best agricultural implements.

Efforts to promote industry during the antebellum years were inconsequential, and the large majority of industrial exhibits were the products of local artisans. Such was to be expected in a state whose economy was overwhelmingly agricultural. The few industrial exhibits that were mounted did, however, accurately reflect the industrial development of the state during that era, although they did little to directly encourage it. Once the fair was reestablished after the Civil War, it succeeded in improving its industrial exhibits, largely because the period coincided with the movement for an industrialized New South and the beginnings of an "industrial revolution" in the state. While the number of exhibits from tobacco, textile, and fertilizer firms increased, a large majority of the state's industrial firms eschewed the fair. Such exhibits as were mounted, however, acquainted North Carolinians with the state's increasing industrialization and caused them to take pride in this development.

During the last two decades of the nineteenth century, the fair worked closely with newly established state agricultural agencies to educate North Carolinians about better agricultural practices. The North Carolina Department of Agriculture employed the fair in its attempts to reach large numbers of the state's farmers and held several farmers' institutes at the fair. The College of Agriculture and Mechanical Arts, the Department of Agriculture's Experiment Station, and the Farmers' Alliance likewise employed the fair as a means of encouraging reforms. The fact that those organizations and institutions relied upon the fair in carrying out their work suggests the significance attached to the fair by the agricultural reform movement. The state fair served as a coordinator for the various phases of the movement and helped to bring reforms to all sections of the state. In 1896 J. W. Carter of the *Richmond Times* characterized the North Carolina State Fair as "one of the best fairs I have seen in the South."

During the twentieth century, the state fair underwent significant changes, not the least of which was the ever-increasing importance of the midway and staged entertainment. The modern midway was fairly well evolved by the turn of the twentieth century, and its significance grew during the 1930s, when North Carolinians caught in a savage depression sought a good time and George Hamid, one of America's most remarkable showmen, ran the fair. Hamid's influence enhanced the appeal of circus acts, revues, and thrill shows to grandstand audiences. The heyday of the carnival midway, with its shows, games of skill, and concessions, followed the Second World War and is exemplified by the long run of the Strates Shows on the fair midway. By the late 1970s a variety of factors, including changing social mores and rapid developments in entertainment technology, beginning with television, resulted in changes on the midway. Rides replaced shows as the main attractions, but games of skill and concessions continued to appeal to an increasingly urban fair crowd.

Grandstand shows, on the other hand, lost their appeal, and the fair abandoned first harness racing and then auto stunt shows, staples of the post–World War II fairs. The dedication of Dorton Arena in 1953 gave the fair the opportunity to showcase nationally known musicians, comedians, and other entertainers, who performed afternoon and night for appreciative audiences. To some extent, the success of such acts in an indoor venue contributed to the loss of appeal and gradual decline of the old grandstand shows.

The post–World War II era saw the rise of another form of exhibit at the fair, partly educational, partly nostalgic. Such exhibits appealed to the desire of the state's increasingly urban and suburban population to remain in touch with its agricultural heritage. The folk festival, the Heritage Circle exhibits, and the "Village of Yesteryear" informed rural and suburban folks about the life their forebears had lived in the not-so-distant past before electricity, interstate highways, computers, and the information age. While such displays invariably stressed the positive aspects of life in North Carolina's rural past, they also conveyed a sense of the demanding physical labor and long working hours required to sustain a family farm. The appeal of such nostalgic exhibits, now among the most popular with fairgoers, is likely to increase as fewer North Carolinians retain any personal connection to agriculture and as agriculture increasingly comes to resemble other forms of industrial production.

While the nineteenth-century state fair had begun to feature exhibits by the North Carolina Department of Agriculture and the North Carolina College of Agriculture and Mechanical Arts, it was in the twentieth century, especially after the fair became a state agency in 1928, that familiarizing millions of North Carolinians with the state's basic economic development strategy became an essential feature of the fair. The fair performed that role effectively, emphasizing several basic economic concepts to fair audiences over a period of years. The need for the state's farmers to adopt the latest scientific agricultural techniques remained the fair's basic message, and exhibits stressed the importance of diversifying the state's agricultural productivity and placing less reliance upon the traditional cash

crops of cotton and tobacco. Exhibits by a variety of state agencies and professional organizations likewise proclaimed the need to diversify the state's economy through industrialization and a better-educated citizenry. In championing industrialization and

An aerialist performs on a trapeze suspended from a helicopter above the 1984 fair. Photo courtesy of NCDA&CS.

became more significant, a trend that rapidly accelerated during the years of prosperity that followed the war. During that period the fair's significance in introducing the state's farmers to the new agricultural implements and machinery required of the modern farm likewise reached its height, before declining by the last quarter of the century as other means of display and distribution replaced the fair as a commercial venue. The fair's emphasis on the marketing of farm products, especially livestock, increased markedly once it became a state agency, especially after World War II. Unlike the commercial value of the fair in the realm of agricultural machinery and household appliances, which began to wane in the late 1960s, the fair's function as an important livestock market has continued into the twenty-first century.

Perhaps the most significant development of the modern fair is not the October event that most North Carolinians think of as *the* state fair but the year-round use of the fair's facilities. With the construction of Dorton Arena in the early 1950s; the Gov. Scott, Graham, and Gov. Holshouser Buildings in the 1970s; and the Gov. Hunt Horse Complex in the early 1980s, the fair gained the ability to serve as a major convention and exposition center. Presently, millions each year attend a variety of trade shows, conventions, professional meetings, horse shows, and other functions, including the traditional weekend flea market, all of which are held in those facilities from November through September. In truth, the state fair is now a large, complex, year-round business, run by a professional staff, that brings millions of dollars into the state and is an important part of the economy of Raleigh, Wake County, and the surrounding region. In fact, the fair has become so successful as a year-round operation that it has outgrown its current facilities. No major new exhibit space, badly needed if the fair is to compete with other southeastern convention and exposition centers in the future, has been constructed since the opening of the Gov. Hunt Horse Complex two decades ago.

Whatever the future of the fair as a year-round convention and exposition venue, the future of the October North Carolina State Fair, the annual extravaganza that millions of North Carolinians know and love, seems secure. The fair will continue to focus on

educational progress, however, fair management realized that the fair could educate the public about those concerns but could not directly affect the course of their development.

The fair's role as a commercial venue, an opportunity for the state's residents, and especially its rural families, to become acquainted with the seemingly endless array of products churned out by the incredible American industrial plant, was evident by the early years of the twentieth century. With the electrification of the state's rural areas prior to the Second World War, that aspect of the fair

A full day spent exploring the attractions of the state fair can be an exhausting experience. Tired fairgoers take a break at the 1984 fair. Photo courtesy of NCDA&CS.

the role of agriculture in North Carolina's economy and to educate the state's urban and suburban population about life on its farms and in its rural communities. It will continue to be the most significant celebration of the state's agricultural heritage, as well as a means of educating the state's nonfarm population about that heritage. As a state agency, it will continue to convey the state's basic economic development policies to the hundreds of thousands who attend each year. Its midway attractions, no matter which company operates them, will continue to thrill the crowds who come anticipating a break from life's routine and some pleasant excitement that can be enjoyed with friends and family. At 150 years of age, the North Carolina State Fair is stronger than ever, a grand institution, and an essential ingredient of the North Carolina experience.

Epilogue

On June 6, 2003, Meg Scott Phipps, commissioner of the North Carolina Department of Agriculture and Consumer Services, resigned her office. Controversy surrounding campaign finances and departmental contract awards prompted that move.

To replace Phipps, Governor Michael F. Easley appointed W. Britt Cobb Jr. as interim commissioner for the Department of Agriculture and Consumer Services. At press time, a permanent commissioner had not been named.

Cobb brings a record of thirty years of employment with the department, having served as assistant director of marketing since 1991. In appointing Cobb, Governor Easley observed that "His knowledge and understanding of the agriculture industry and the inner workings of this agency will ensure that the department operates smoothly and efficiently to serve the people of this state effectively throughout this transition period."

One of Cobb's first tasks will be to oversee preparations for the 150th anniversary of the North Carolina State Fair. It is no small task. Planning for the fair is a year-round job, and just to prepare to open each October requires an investment of well over $2 million and the work of literally hundreds of people. Efforts to insure a successful 2003 fair will be aided by the fact that the state fair is a robust institution, virtually self-sufficient financially, managed by a highly competent professional staff, and highly regarded by North Carolinians. Yet the state fair faces substantial challenges in the immediate future, some of which fair manager Wesley Wyatt outlined in a recent interview.

One of the most pressing, Wyatt said, became evident with the terrorists' attacks against America on September 11, 2001, which instantly made security an overriding concern. With October around the corner, the fair had to scramble to increase security. Management immediately initiated a procedure to search bags that gatekeepers thought should be searched and recommended that people not bring large bags to the fairgrounds. Insuring that those bent on doing harm to the fair-going public not penetrate the fairgrounds remains a vital concern, and is likely to remain so for the immediate future. The terrorists' attacks also heightened concerns about food safety because of the possibility that contaminated food or water might be used to attack the large crowds that attend each year. The attacks also increased concerns about animal health issues, which had already been heightened by recent foot and mouth disease outbreaks in Europe. Now and in the future the fair must take extra precautions to insure that there is no deliberate effort to expose livestock on exhibit to foreign animal diseases. This is an especially difficult task, as young people have their animals at the state fair for the public to view and to touch. The fair must also be concerned about the health of the public as well as that of the animals, because of the large numbers of people coming in close contact with them.

Space, according to Wyatt, also remains a crucial need of the fair. The fair badly needs better facilities to meet the constantly increasing demands of individuals, businesses, and organizations seeking to lease space. Space is needed not only for the ten-day fair in October, but also for a variety of potential exhibitors all year. While existing exhibit space can perhaps be better organized, Wyatt said, current demand, both at the ten-day fair and

Dorton Arena and the waterfall, two enduring symbols of the state fair. Photo (2002) courtesy of the author.

during the remainder of the year, make it clear that the fair's exhibit space needs to be enlarged. The fair's management has proposed state funding for the construction of a new multipurpose building, which would definitely help meet this need. Fair management also hopes to transform the current infield of the grandstands' racetrack, making it available for the relocation of midway attractions during the ten-day fair and for other uses during the year.

Created in the nineteenth century when the overwhelming majority of North Carolinians lived on farms, the state fair now serves North Carolinians who reside primarily in cities and suburbs. While agribusiness remains a potent force in the state's economy, very few people actually till the soil or tend to flocks or herds. It is true that North Carolinians come to the autumn spectacle for the carnival, for the rides and games, for the candy apples and cotton candy, for the recording stars who perform in Dorton Arena, and for the tractor pulls and strolling performers. But that is only part of the story. Those types of events can be staged any time, at any venue. The North Carolina

State Fair continues to be about 4-H Club members readying livestock for exhibit; about prize dairy and beef cattle; about the best displays of cotton, corn, tobacco, and other crops; about cured hams and perfect yams; about judges tasting home-canned fruits, vegetables, pickles, and relishes; about home-baked breads, pies, and cakes; about displays of homemade clothing and hand-stitched quilts; and about exhibits carefully crafted by residents of rural communities.

The North Carolina State Fair remains the one institution at which people from across the state gather to celebrate their agricultural heritage. It is where North Carolinians remind themselves and their children of how and by whom the abundance of food and fiber that sustains the complex and incredibly wealthy society in which they reside is produced. In serving that purpose, the fair remains faithful to the ideals of those North Carolinians who, 150 years ago, determined that North Carolina needed a state fair. No doubt other methods could have achieved many of the same goals. In the future, perhaps, some fabulous website could portray the sights and sounds of every conceivable form of agricultural production and display a profusely illustrated, encyclopedic account of all aspects of the state's agricultural history. But it could never capture the sheer fun of attending the fair, nor create the memories that for many are part of the experience of being a North Carolinian. For a century and a half, the North Carolina State Fair has contributed to the economic and social life of the state, and it remains one of the state's most significant and best-loved institutions. Now, as in the past, when October comes and the days grow shorter, when a southward-shifting sun warms the twilight sky, when the night air is again crisp and clear, North Carolinians know that it is once more time for that annual celebration of their agricultural heritage. They know it is state fair time, and most, one suspects, hope that it will always be so.

Essay on Sources

Surprisingly, there is little written on the North Carolina State Fair, an institution attended by literally millions of people during its 150-year history. The state fair, an institution that has helped to shape what it means to be a North Carolinian, is part of the shared memories of millions of present-day North Carolinians. Fortunately, its long history of educating and entertaining the citizens of North Carolina can be compiled from records that exist in the state's archival collections.

This work is based on a wide variety of both primary and secondary sources, including publications of the North Carolina State Agricultural Society and the North Carolina Department of Agriculture, archival collections, newspapers, journal articles, various Web sites, autobiographies, personal interviews, and several historical studies. This essay, not intended as a comprehensive bibliography, is instead designed as a guide for those who might wish to further study the history of the North Carolina State Fair.

Three major sources provide an overview of the growth and development of the fair. The *North Carolina State Fair Premium Lists*, published annually by the Agricultural Society until 1925 and after 1928 by the North Carolina Department of Agriculture, are perhaps the single most important source. They can be found in the North Carolina State Archives and in the North Carolina Collection of the University of North Carolina at Chapel Hill. Also to be found in the State Archives is the *Agricultural Review*, published bimonthly by the Department of Agriculture since 1928. The *Program of the North Carolina State Fair*, also published annually by the Department of Agriculture, is an indispensable source, but complete runs are not available. Copies of programs for most fairs since the Second World War can be found in either the State Archives or in the files of the state fair at the Department of Agriculture. The 1953 program *North Carolina State Fair Centennial, 1853–1953*, also available at the North Carolina Collection, provides an excellent summary of the history of the fair up to that date. In addition to those three sources, a variety of newsletters concerning the fair during the 1960s and 1970s can be found in the files of the Department of Agriculture. The State Archives contains a variety of materials on the fair, some of the most useful from the papers of W. Kerr Scott. As would be expected of the major newspaper in the capital city, the Raleigh *News and Observer* provides excellent material on the twentieth-century fair, especially through the 1970s. The material on the nineteenth-century fair is largely taken from Melton McLaurin, "The

North Carolina State Fair, 1853–1899" (master's thesis, East Carolina College, 1963).

James West Potter, who is currently developing a history of professional wrestling in the Raleigh area, provided information about that subject. Derek Nelson, *The American State Fair* (Osceola, Wis.: MBI Publishing, 1999) is richly illustrated and contains an excellent overview of the origins and development of a uniquely American institution. Two of the foremost attractions on the state fair midway are chronicled in Al W. Stencell, *Girl Show: Into the Canvas World of Bump and Grind* (Toronto, Canada: ECW Press, 1999), a profusely illustrated and well-written account of a vanished aspect of the carnival world, and Robert Bogdan, *Freak Show, Presenting Human Oddities for Amusement and Profit* (Chicago: University of Chicago Press, 1988). Details on the 1893 switchback railway accident are contained in Katherine Batts Salley, ed., *Life at Saint Mary's* (Chapel Hill: University of North Carolina Press, 1942).

Autobiographical and biographical works provide significant insights about major figures in the history of the fair. Much of the material on George Hamid's impact upon the fair is taken from George A. Hamid, *Circus* (New York: Stirling Publications, 1950). Jim Graham's autobiography, *The Sodfather, A Friend of Agriculture*, is available on a website maintained by North Carolina State University at *www.lib.ncsu.edu/archives/exhibits/ sodfather*. Material on the lives of both James E. and E. James Strates is available on the Strates Shows website: *www.strates.com*. Loyal Jones's *Minstrel of the Appalachians: The Story of Bascom Lamar Lunsford* (Boone: Appalachian Consortium Press, 1984) provides details of the life of the founder of the State Fair Folk Festival.

Among journal articles consulted was Jim L. Sumner, "Presidential Visits to the North Carolina State Fair," *Carolina Comments* 36 (May 1988): 74–81, an excellent overview of that subject. Material on the fair during the First World War is found in a collection of documents reprinted in "War Camp Community Service," *North Carolina Historical Review* 1 (October 1924): 412–448; and an untitled article on Camp Polk after the First World War by Frederick A. Olds can be found in the January 17, 1919, issue of *Orphans Friend and Masonic Journal*, pp. 1–6.

William S. Powell, *North Carolina through Four Centuries* (Chapel Hill: University of North Carolina Press, 1989) provided the historical backdrop for the fair's development, as did James Vickers, *Raleigh, City of Oaks: An Illustrated History* (Woodland Hills, Calif.: Windsor Publications,

1982), and Elizabeth Reid Murray, *Wake, Capital County of North Carolina*, Vol. I: *Prehistory through Centennial* (Raleigh: Capital County Publishing Company, 1983). Two works were relied upon to chronicle the changes that transformed southern agriculture in the middle of the twentieth century: Pete Daniels, *Breaking the Land: The Transformation of Cotton, Tobacco and Rice Cultures since 1880* (Urbana: University of Illinois Press, 1985), and Jack Temple Kirby, *Rural Worlds Lost: The American South, 1920–1960* (Baton Rouge: Louisiana State University Press, 1987). John Haley, *Charles N. Hunter and Race Relations in North Carolina* (Chapel Hill: University of North Carolina Press, 1987) provided details on the Negro State Fair.

Finally, a number of individuals consented to provide the author with personal interviews, and their willingness to share their knowledge of and insights about the state fair is deeply appreciated. They include James Graham, then North Carolina commissioner of agriculture, interviewed December 7, 2000; E. James Strates, president of Strates Shows, Inc., interviewed October 20, 2000; Wesley Wyatt, manager of the North Carolina State Fair, interviewed May 17, 2002; Robert Zimmerman, president of Southern Shows, Inc., interviewed July 11, 2002; and Meg Scott Phipps, former North Carolina commissioner of agriculture, interviewed August 6, 2002.

Index

ML   7/05